MEGA-FUN FRACTIONS

by Marcia Miller and Martin Lee

SCHOLASTIC
PROFESSIONAL BOOKS

NEW YORK · TORONTO · LONDON · AUCKLAND · SYDNEY
MEXICO CITY · NEW DELHI · HONG KONG · BUENOS AIRES

With love to Daniel and Joshua Brandes

Cover design by Maria Lilja
Interior design by Solutions by Design, Inc.
Cover and interior illustrations by Michael Moran

ISBN: 0-439-28844-4

CONTENTS

ABOUT THIS BOOK

NCTM Standards

In the 2000 edition of *Principles and Standards for School Mathematics*, the National Council of Teachers of Mathematics (NCTM) states that students in grades 3–5 should:

✴ Develop understanding of fractions as parts of unit wholes, as parts of a collection, as locations on number lines, and as divisions of whole numbers

✴ Use models, benchmarks, and equivalent forms to judge the size of fractions

✴ Develop and use strategies to estimate computations involving fractions in situations relevant to students' experience

✴ Use visual models, benchmarks, and equivalent forms to add and subtract commonly used fractions

Our Goal

We have written *Mega-Fun Fractions* to provide in one resource a variety of ways for you to immerse your students in fraction concepts. All activities address one or more of the NCTM fraction standards listed above. The range of fraction lessons includes hands-on explorations and activities that invoke problem solving, reasoning and proving, communicating, connecting, and representing fractions. Cross-curricular activities link fractions to language arts, music, science, art, and social studies.

We hope that as you use the ideas in this book in your classroom, your students will develop a deeper understanding of fractions and become more comfortable with this strand of the mathematics curriculum. Our goal is to foster a strong conceptual understanding that, we believe, will lead to greater ease in working with fractions and rational numbers at more advanced levels.

The Format

Mega-Fun Fractions offers activities written directly to the student as well as guided plans to help you present activities to your whole class, to small groups, or to individuals. Each lesson begins with a question you may pose to students, a learning objective, a list of necessary materials, sequenced steps to follow (The Plan), and several follow-up ideas. The back of the book has three pages of Fraction Quickies—additional ideas presented in an abbreviated manner. You will also find helpful reproducibles and a Fraction Self-Evaluation Form that students complete. Answers appear at the end of the book.

Teacher Tips

✴ Refer to the chart on page 6 to find activities that focus on a particular skill.

✴ Feel free to work through the book in any order that suits you.

✴ You may find that some activities are too advanced or basic for your class. Adapt tasks to suit your students' needs.

✴ You may choose to use the tasks in this book as full lessons, warm-ups, homework assignments, math corner activities, group projects, informal assessments, or portfolio assessments.

✴ Determine the best grouping to suit your teaching style, as well as the learning styles and levels of your students. Invite students to work individually, in pairs, in small groups, or as an entire class. Be sure to allow time for sharing and comparing.

✴ Encourage sharing, discussing, analyzing, and summarizing of students' findings.

✴ Duplicate and distribute the fraction strips (page 84) to provide students with a ready reference for comparing and ordering fractions. Have students color each strip a different color.

✴ The Prepare to Share feature on many student pages stimulates students to plan what they wish to say when you summarize the activity. You may wish to ask students to record their responses in a math log.

Skills Matrix

	Fraction Concepts	Visual Estimation	Fractions of a Region	Fractions of a Set	Equivalent Fractions	Comparing and Ordering Fractions	Rounding and Estimating	Interpreting Charts and Data	Measurement and Money	Mixed Numbers	Adding Fractions	Subtracting Fractions	Multiplying Fractions
Halfness, p. 7	X	X	X										
Fill All Four, p. 9	X		X										
What's in Petal, Mississippi? p. 9	X	X	X										
What's Left? p. 10	X	X	X										
Fractions up a Tree, p. 10	X	X	X										
Colorful Regions, p. 15	X		X										
Fraction Chart Puzzle, p. 16	X												
Part Art, p. 18	X		X										
Collaborative Quilts, p. 19	X		X										
Pattern Block Proofs, p. 20	X		X		X								
Shaded Shapes, p. 23	X		X		X								
Stand Up for Fractions, p. 25	X			X									
Class Fraction Wall, p. 26	X			X				X					
Picture These Fractions, p. 27	X			X									
Mixing Snack Mix, p. 28	X			X									
Fraction Message, p. 28	X			X									
The Language of Fractions, p. 28	X												
Fraction Dictation, p. 32	X									X			
Fractions and Egg Cartons, p. 33	X				X								
Badge Buddies, p. 34	X				X								
Equivalent Fraction Concentration, p. 35	X				X								
Fractions Bingo—Times Two, p. 36	X				X		X			X			
Fraction Poems, p. 39	X												
Sharing Fraction Pie, p. 41	X					X							
Fraction Fill 'Em Up, p. 42	X					X							
Prove It! p. 43	X					X							
Fraction War, p. 44	X				X	X							
Fraction Rollers, Part 1, p. 45	X					X							
That's an Order! p. 46	X					X							
Fractions of a Day, p. 47	X							X	X				
Time for Fractions, p. 50	X								X				
Coining Fractions, p. 51	X								X				
Funny Money, p. 52	X								X				
Fractions and Ages, p. 53	X						X	X	X				
Fraction Rollers, Part 2, p. 54	X					X				X			
Noting Fractions, p. 55	X							X		X			
A Head for Fractions, p. 57	X						X			X			
Fractions in Ancient Egypt, p. 58	X									X			
Flicker Fraction Sums, p. 60	X									X			
Fraction Add-Up, p. 63	X								X	X	X	X	
Sums on a Roll, p. 65	X					X				X	X		
Fraction Path Puzzles, p. 66	X										X	X	
Fraction Magic Figures, p. 68	X										X	X	
Parts of Parts, p. 70	X												X
Roll, Round, and Record, p. 72	X						X			X			X
Fractions and Calories, p. 74	X							X	X	X			
Fraction Stories, p. 76	X										X	X	X
Fraction Scavenger Hunt, p. 77	X	X							X				
Only One-Third Agreed That..., p. 79	X			X				X					
Fractions Every Day, p. 80	X												

HALFNESS

Can you open a book to the halfway page?
Can you pour half a glass of water?
Can you walk halfway to the classroom door
from your seat?

GOAL: Students use their visual estimation skills to identify half of a region.

MATERIALS: student page 8, various books, water glasses, dried beans or water, coins or counters, string, pencils, various measurement tools (rulers, tape measures, scales, and so on)

THE PLAN

1. Begin by brainstorming with students everyday situations in which people make a quick visual estimate of half. For example:
   - ✸ I'll take half a piece, please.
   - ✸ Fill it up about halfway.
   - ✸ Hang it up about halfway between here and the door.
   - ✸ The book goes in the middle of the top shelf.

2. Continue by having students suggest other examples that involve estimating about half of a space, a distance, or a quantity.

3. Divide the class into pairs or small groups. Give each group a set of materials. Then direct groups to tackle the visual estimation activities on page 8. Have students try each task more than once, and note whether their estimates of half improve or stay about the same.

4. As students work, guide them to figure out ways to use suitable measurement tools to verify how close to half their estimates actually are.

TEACHING TIPS

- ✸ Have students compare their estimates and strategies with those of other groups.
- ✸ Invite discussion of students' strengths and weaknesses in visual estimation.
- ✸ Ask students to describe orally or write a description of how and why their estimates changed with subsequent repetitions.
- ✸ Invite students to create other tasks in which they visually estimate half of something.
- ✸ Challenge students to visually estimate "half again" of a distance or quantity.

HALFNESS

Your senses can help with fractions.
It's great if you can visually estimate half.

Work in pairs or small groups. Try each task three times to sharpen your fraction estimating skills. Judge how close your estimates are. Do your estimates get better the second and third time?

⊛ Choose a book of any thickness. Open it to its middle page.

⊛ Fill a glass halfway with beans or water.

⊛ Place two coins or counters a distance apart on the floor or on a table. Then place another counter halfway between them.

⊛ Have two classmates stand a reasonable distance apart. Stand halfway between them.

⊛ Choose something within the classroom, at a distance from you. Walk halfway there. Mark where you started and where you stopped.

⊛ Use string to make a large, closed irregular shape on a desktop. Place a row of pencils across it to divide the shape in half.

⊛ Have a classmate stand tall. Show where half his or her height is.

PREPARE TO SHARE

How did you check your visual estimates?
Did your estimates get better as you went along? Explain.
Which kind of visual estimation was the hardest for you? The easiest?

FILL ALL FOUR

Play a two-person game that involves filling regions of a hexagon.

GOAL: Students use visual and spatial reasoning skills to fill hexagonal regions.

MATERIALS: student page 11, number cube (p. 85), pattern blocks (p. 86)

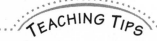

- ✸ Go over the rules of the game with students. Have each student choose a partner. Provide each student with student page 11 and enough pattern blocks to play. As an alternative, students might play in two-person teams.

- ✸ Give each student a blank number cube. Guide them as they fill in the cube according to the directions on page 11.

- ✸ Challenge students to name the fractional amount of the hexagons that are filled at any given point in the game.

- ✸ Invite students to play several times to sharpen their strategies for mastering the game.

- ✸ Have students compare their game-playing strategies.

WHAT'S IN PETAL, MISSISSIPPI?

Identify fractions of regions to answer a riddle.

GOAL: Students name fractional parts of a region.

MATERIALS: student page 12

- ✸ Duplicate and distribute copies of page 12 to each student. Explain that if students correctly name the fractional part of each balloon and fill in the correct number in the coded answer below, they will figure out the answer to the question about Petal, Mississippi.

- ✸ Invite students to develop variations of this puzzle for classmates.

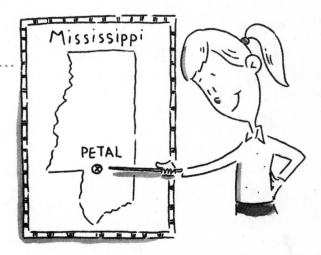

What's Left?

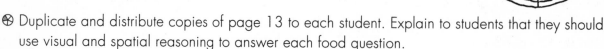

Use visual estimation to describe fractional parts.

GOAL: Students apply visual and spatial reasoning to name fractional parts of foods.

MATERIALS: student page 13

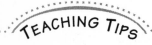

TEACHING TIPS

⊛ Duplicate and distribute copies of page 13 to each student. Explain to students that they should use visual and spatial reasoning to answer each food question.

⊛ Invite students to develop variations of this page for classmates.

Fractions Up a Tree

Use visual and spatial estimation to follow fraction clues.

GOAL: Students draw a hole that represents where a woodpecker pierced a tree.

MATERIALS: student page 14

TEACHING TIPS

⊛ Duplicate and distribute copies of page 14 to each student. Explain that students will use visual and spatial reasoning to draw the position of the woodpecker hole in each tree.

⊛ Invite students to develop variations of this task for classmates.

⊛ Translate this exercise into three dimensions by having students do similar tasks with drinking straws, cardboard tubes, interlocking blocks, or clay cylinders.

FILL ALL FOUR

Play a two-person game. Each player needs this game board, pattern blocks, and a number cube labeled so that one face says $\frac{1}{2}$, two faces say $\frac{1}{3}$, and three faces say $\frac{1}{6}$.

The object of the game is to fill the four hexagons with fractional parts. Each hexagon = 1 whole.

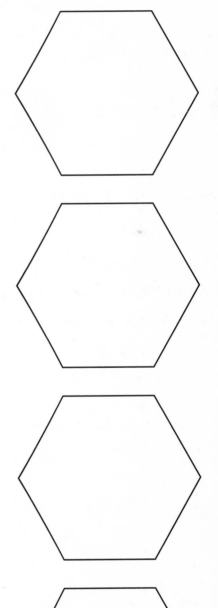

RULES

1. Before you play, find out which pattern block fills $\frac{1}{2}$ the hexagon. Determine which pattern block fills $\frac{1}{3}$ of the hexagon, and which fills $\frac{1}{6}$ of it. Use these pattern blocks to play.

2. Decide who goes first. That player rolls the fraction number cube and takes a pattern block for the fraction shown. The player puts that block inside any hexagon where it will fit. Once a piece has been placed, it may not be moved.

3. The next player rolls, takes the corresponding pattern block, and puts it anywhere it will fit.

4. In turn, players keep rolling and adding pattern blocks to the hexagons to fill all four of them.

5. Any player who rolls a fraction that does not fit in an available space may roll again. If the second roll does not fit either, that player's turn ends.

6. The first player to fill all four hexagons wins.

STRATEGY HINT
It's okay to mix halves, thirds, and sixths in one hexagon as long as each piece clearly fits.

WHAT'S IN PETAL, MISSISSIPPI?

Since 1976 the town of Petal, Mississippi, has had something that no other town has. What is it?

To find out, look at the balloons. Above each fraction, write the letter of the balloon whose shaded area shows that fractional amount.

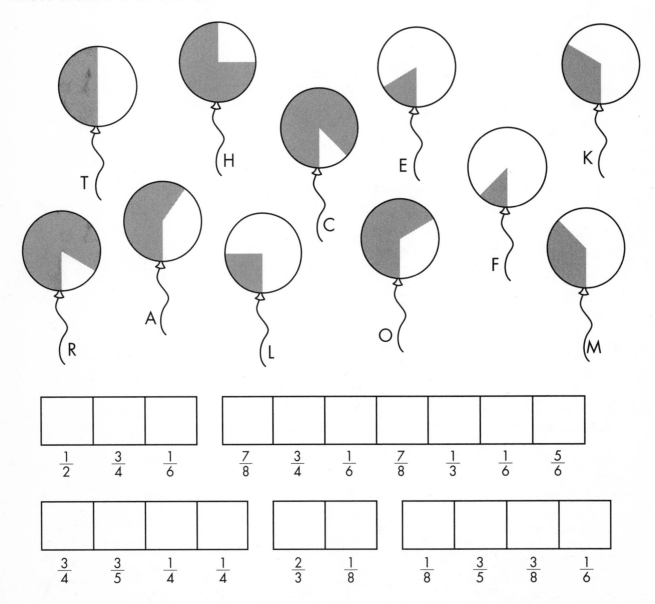

$\frac{1}{2}$	$\frac{3}{4}$	$\frac{1}{6}$

$\frac{7}{8}$	$\frac{3}{4}$	$\frac{1}{6}$	$\frac{7}{8}$	$\frac{1}{3}$	$\frac{1}{6}$	$\frac{5}{6}$

$\frac{3}{4}$	$\frac{3}{5}$	$\frac{1}{4}$	$\frac{1}{4}$

$\frac{2}{3}$	$\frac{1}{8}$

$\frac{1}{8}$	$\frac{3}{5}$	$\frac{3}{8}$	$\frac{1}{6}$

PREPARE TO SHARE ....

What would you expect to see at this unique place in Petal, Mississippi?

Mega-Fun Fractions ⊛ Scholastic Professional Books

WHAT'S LEFT?

Estimate to figure out about how much food is left. Then write the fraction that is closest to that amount.

Norm's sandwich

1 Norm didn't finish his sandwich. About what fraction of it remains? Write $\frac{1}{4}$, $\frac{1}{2}$, or $\frac{3}{4}$.

Norm left about _____.

What's left

2 Some people didn't finish their pizzas. About what fraction of each pizza is left? Write $\frac{1}{4}$, $\frac{1}{2}$, or $\frac{3}{4}$.

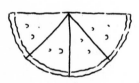 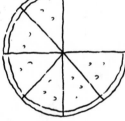

a. _____   b. _____   c. _____

3 There is some juice left in each glass. About how full is each glass of juice? Write $\frac{1}{6}$, $\frac{1}{3}$, $\frac{2}{3}$, or $\frac{5}{6}$.

a. _____   b. _____   c. _____

4 About how much of each bread loaf is left? Write $\frac{1}{8}$, $\frac{3}{8}$, $\frac{5}{8}$, or $\frac{7}{8}$.

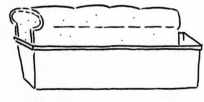 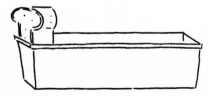 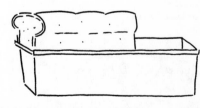

a. _____   b. _____   c. _____

FRACTIONS UP A TREE

Hungry woodpeckers pecked each of these trees. Use fraction clues to figure out how far up the trunk they pecked. Draw each hole about where it should be.

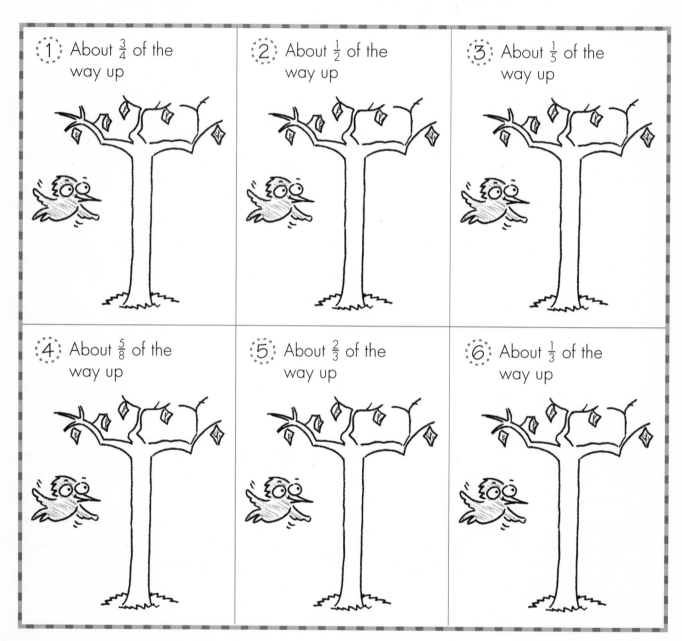

1: About $\frac{3}{4}$ of the way up

2: About $\frac{1}{2}$ of the way up

3: About $\frac{1}{5}$ of the way up

4: About $\frac{5}{8}$ of the way up

5: About $\frac{2}{3}$ of the way up

6: About $\frac{1}{3}$ of the way up

PREPARE TO SHARE

How did you decide where to draw each hole? Explain.

COLORFUL REGIONS

Can you color a rectangle on dot paper to represent a given fraction?

GOAL: Students practice naming, writing, and drawing fractional parts of a region.

MATERIALS: scrap paper, dot paper (p. 87), rulers, crayons

THE PLAN

1. Divide the class into pairs. Ask each partner to write five different fractions (less than 1) on a piece of scrap paper. They should write some fractions in number form and others in words. Have partners swap their papers.

2. Distribute dot paper and rulers to each student.

3. Have students draw five separate rectangles on their dot paper. Explain that each rectangle drawn represents one whole and that each may be the same size and shape or different. Students will use these rectangles to illustrate the fractions their partners wrote for them.

4. Have students color regions of the five rectangles to show the fractions written. Students may use the rulers to outline the regions and the colored parts.

5. Ask partners to share their drawings and discuss them. Encourage them to work out any discrepancies.

FOLLOW-UP

⊛ Have each pair of students swap their drawings with another pair of students. Each pair should then identify the fractions shown by the colored parts.

⊛ Extend by having them identify two fractions to be colored in the same figure. For example, a student can ask another to color a rectangle $\frac{1}{5}$ blue and $\frac{3}{5}$ yellow. If students are familiar with the concept of common multiples, you may wish to extend the activity by suggesting that partners come up with fractions that have different denominators.

⊛ Challenge students by having them color fractional parts within irregular polygons. It may be useful first to work one together on the overhead projector or chalkboard.

FRACTION CHART PUZZLE

Can you solve a fraction puzzle?

GOAL: Students assemble a schematic chart that identifies the parts of a fraction.

MATERIALS: student page 17, scissors, glue or tape, plain paper

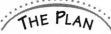
THE PLAN

1. Begin by reviewing the names of the parts of a fraction (numerator, denominator, fraction bar). Reinforce the role of each part in giving meaning to the fraction.

2. Duplicate and distribute copies of page 17. Explain that the puzzle contains a set of scrambled parts of a fraction as well as phrases that describe the function of each part.

3. Provide scissors, glue or tape, and sheets of plain paper. Challenge students to cut out the parts of the fraction puzzle and the descriptive phrases and then reassemble them in proper order to represent a fraction.

4. Display the completed fraction puzzles, or have students keep them in their math notebooks to use as a reference tool.

FOLLOW-UP

⊗ Extend this activity by creating similar puzzles to represent mixed numbers.

⊗ Create an interactive fraction chart using two library pockets fastened to posterboard. Students can form an endless variety of fractions by slipping number cards into the pockets.

FRACTION CHART PUZZLE

How can you unscramble these parts to form a fraction model?

Cut out the seven pieces along the dotted lines. On another sheet of paper, glue them in position to form a labeled fraction model.

how many to consider

denominator

numerator

how many in all

PREPARE TO SHARE ....

How might you use the completed model to teach others about fractions? Explain on a separate sheet of paper.

PART ART

In how many ways can you use fractions to describe a geometric design?

GOAL: Students use fractions to describe geometric designs.

MATERIALS: centimeter grid paper (p. 88), crayons, ruler

THE PLAN

1. On an overhead projector, chart paper, or chalkboard, create the design shown below.

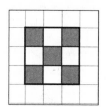

2. Ask students to describe the design by identifying its fractional parts. Students may say, for example, that the design is $\frac{5}{9}$ black. Others may say that it is $\frac{4}{9}$ white. Some may even say that the number of white squares is $\frac{4}{5}$ of the number of black squares. Record students' suggestions.

3. Reproduce and distribute sheets of centimeter grid paper and provide crayons. Tell students to outline a 4 x 4 square on the grid paper. Have them color the small squares within the figure to create a geometric design. Invite them to use as many colors as they wish.

4. Have students write fractions that describe their designs. Encourage them to do so in as many ways as they can. Guide them to see that as they increase the number of colors used in a design, they increase the number of ways in which they can describe that design with fractions.

5. Next, ask students to repeat the process with another 4 x 4 square. This time, have them record their descriptions on separate paper. Then have students swap their designs (but not the descriptions) with a partner. The partners write fractions to describe the design in as many ways as they can. When they are done, partners compare descriptions of each other's design.

6. Have pairs repeat this process using 5 x 5 or 6 x 6 squares.

FOLLOW-UP

⊗ Post students' designs with the fractional descriptions.

⊗ Challenge pairs to envision a simple geometric design on a 4 x 4 square. Then ask each student to give his or her partner a full verbal description of that design, using fractions and directional language (upper left, lower right, and so on). The partner tries to re-create the design described. Have partners swap roles.

COLLABORATIVE QUILTS

How can you use fractions to create and describe unique class quilts?

GOAL: Students apply their understanding of fractional parts to make geometric designs on patches they will combine into one large quilt.

MATERIALS: pictures of various quilts, inch grid paper (p. 89), scissors, crayons or markers, plastic shower curtain or tablecloth to serve as backing for quilt, tape or glue

THE PLAN

1. Display pictures of quilts that reflect a variety of patterns. Ask students to imagine a collaborative class quilt in which each child contributes one or more patches.

2. Reproduce and distribute sheets of inch grid paper and provide crayons. Have students cut the grid paper into 6 x 6 squares so that everyone has an equal-sized patch.

3. Decide with the class on fractional parts to go in the quilt. For example, students can each decorate their patch to represent $\frac{1}{2}$, $\frac{1}{3}$, $\frac{1}{4}$, $\frac{1}{6}$, or $\frac{1}{8}$. Students can decorate their patches in any way they wish, as long as they can describe it in terms of fractions.

4. As students complete their patches, tape or glue the patches onto the shower curtain or tablecloth backing to form a communal quilt you can display. As an alternative, combine the patches to form a quilt on a bulletin board by connecting them with tacks or staples.

FOLLOW-UP

⊗ Invite students to describe the quilt using fractions. Guide them to describe individual patches, certain rows or columns, border patches, or the entire quilt.

⊗ Have the class work together to create a real quilt by sewing together fabric patches. You might focus on cross-curricular themes, such as history, science, geography, or literature. It may help to invite parent volunteers to assist you with this project.

⊗ Challenge students to analyze existing quilts (actual ones or pictures of them) by describing them in fractional terms.

PATTERN BLOCK PROOFS

How do pattern blocks reflect the concept of equivalent fractions?

GOAL: Students use pattern blocks to identify relationships among fractions and to explore the concept of equivalent fractions.

MATERIALS: sets of pattern blocks (p. 86), student pages 21–22

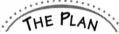 THE PLAN

(1) Distribute pattern blocks to students. Allow time for them to examine and identify the blocks in each set. Have students color the shapes as follows: yellow hexagon, red trapezoid, blue parallelogram [or rhombus], red square, white parallelogram [or rhombus], and green triangle.

(2) Allow students to create designs with pattern blocks to explore how they fit together. Then have them manipulate the pattern blocks to answer questions, such as:

✲ How many green blocks cover a blue block? (*2*)

✲ Which red block covers half a yellow block? (*the trapezoid*)

✲ How many green triangles cover the red trapezoid? (*3*)

✲ How many green triangles cover the yellow hexagon? (*6*)

(3) Have students cover half of a yellow block with the red trapezoid. Also have them cover half of another yellow block with green triangles. Guide students to grasp that since the red trapezoid is $\frac{1}{2}$ of the yellow block, and that since 3 green triangles also cover $\frac{1}{2}$ of the yellow block, then it stands to reason that $\frac{3}{6}$ and $\frac{1}{2}$ are equivalent fractions.

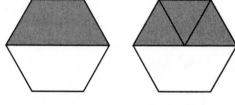

(4) Now ask students to use the pattern blocks to figure out how many blue blocks cover a yellow, and how many green blocks cover a blue. (*3;2*)

Guide students to use these facts to prove that $\frac{1}{3} = \frac{2}{6}$.

(5) Duplicate and distribute the following pages. Ask students to use the figures to answer the questions at the bottom of the page.

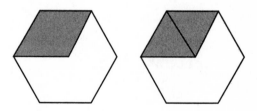

 FOLLOW-UP

✲ Challenge students to use the yellow, blue, and green pattern blocks to prove how many sixths are equivalent to $\frac{2}{3}$. Pose other comparable equivalence questions for students to prove with the pattern blocks.

Name _____ Date _____

PATTERN BLOCK PROOFS

Work with a partner. Use your pattern blocks to make each figure.
Use what you know about the blocks to answer the questions.

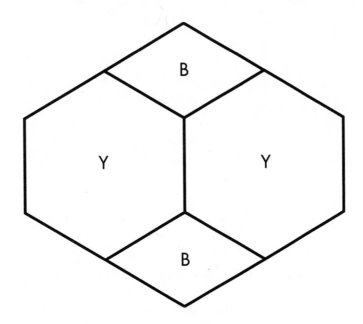

1 What fraction of the whole figure does one blue block cover? _____

    What fraction do two blue blocks cover? _____

2 What fraction does one yellow block cover? _____

    What fraction do three blue blocks cover? _____

3 What fraction of the figure would two green blocks cover? _____

    What fraction would two red blocks cover? _____

Name _____ Date _____

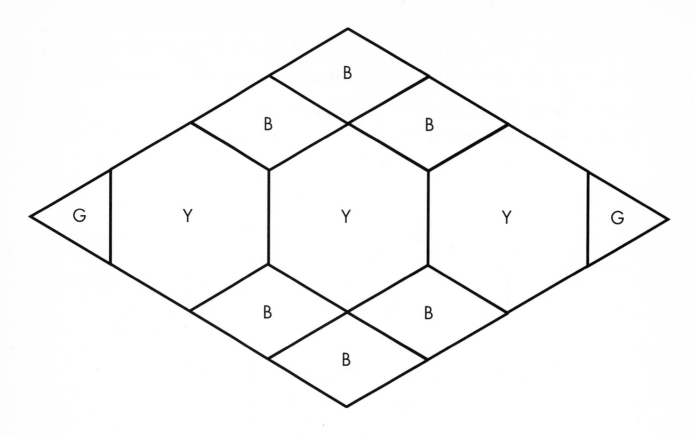

4. What fraction of the whole figure does one green block cover? _____

   What fraction does one yellow block cover? _____

5. What fraction of the whole figure do three blue blocks cover? _____

   What fraction do three yellow blocks cover? _____

6. Which blocks will cover $\frac{1}{2}$ of the figure? Give as many combinations as you can.

   _____

   _____

   _____

   _____

   _____

Mega-Fun Fractions ✵ Scholastic Professional Books

SHADED SHAPES

How can visual reasoning help you shade the same fractional parts of figures in different ways?

GOAL: Students shade portions of figures to explore the concept of equivalent fractions.

MATERIALS: ruler, centimeter grid paper (p. 88), student page 24

THE PLAN

1. On a grid on an overhead projector, draw several 3 x 4 rectangles. Elicit from students different ways to shade the figures to show $\frac{1}{2}$. Let volunteers do the shading.

2. If students don't suggest it, show how to divide the figure in half diagonally. Then, if it has not yet been suggested, show the following ways to shade $\frac{1}{2}$:

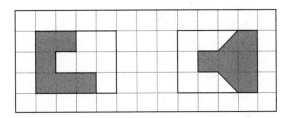

3. Discuss how this shading also represents $\frac{1}{2}$ of the figure. Challenge students to present other ways to shade half of the figure. Have them explain their reasoning.

4. Duplicate and distribute page 24 along with several sheets of centimeter grid paper. Have students work in pairs to show the fractions.

5. Invite pairs to share their sketches with others and to explain and discuss their approaches and results.

6. Have pairs share and discuss their solutions to the Brain Tickler.

FOLLOW-UP

⊗ Challenge students to come up with as many ways as they can to shade the same fraction of the figure shown below.

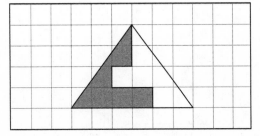

SHADED SHAPES

In how many different ways can you show a given fraction of a region?

Study each figure. See how the shaded part represents the fraction. Then copy the figure onto grid paper. Show as many other ways as you can to shade that same fractional part.

1. The shaded part shows one-half.

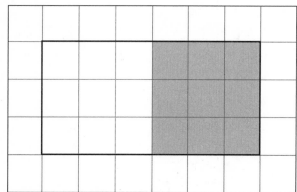

2. The shaded part shows one-half.

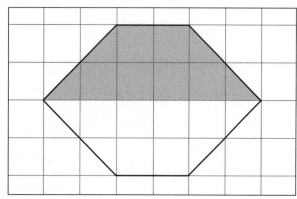

BRAIN TICKLER
The square below is made up of four small squares. Shade half of it so that the unshaded part is also a square.

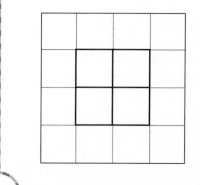

3. The shaded part shows one-third.

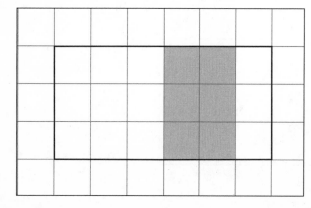

Mega-Fun Fractions ⊛ Scholastic Professional Books

STAND UP FOR FRACTIONS

How can fractions help you describe what your class is wearing today?

GOAL: Students apply their understanding of the concept of fractions of a set to describe features of a group.

MATERIALS: only your students and their clothing!

THE PLAN

1. Record the number of students in class today—boys + girls. Ask, "What fraction of the class is the number of girls? What fraction are boys?" Explain that the total number of students in class is an example of a set, as are the number of boys and girls. Remind students that fractions can be used to describe features of a set. Point out that today they will be describing clothing as the specific feature of sets.

2. List different ways students might identify or categorize the clothing they wear to school. Guide them to suggest a variety of descriptions, from basic categories, such as sweater, skirt, dress, pants, or shoes, to more specific ones, such as long-sleeved top, button-front shirt, T-shirt with writing, jeans that are not blue, sneakers without laces, and so on.

3. Using the list students create, ask all who are wearing one kind of clothing to stand. For instance, say, "If you are wearing a shirt that buttons down the front, please stand." Have a volunteer count the students on their feet and then use a fraction to identify this standing group. For example, if 7 students are wearing button-down shirts and there are 19 students in the class, the fraction $\frac{7}{19}$ identifies the standing group. The numerator is the number of students wearing button-down shirts and the denominator is the total number of students in the class. If it is appropriate for your class, encourage students to express all fractions in simplest form. Have a volunteer record the fractions next to the clothing categories listed.

4. Repeat this procedure for the different garments students listed. Extend the task to include questions like "What fraction of all students is the set of girls wearing jeans?"

5. When all the data has been gathered, have students analyze it. Have them form small groups to carefully examine the fractions together. Ask them to write a summary of the information gathered about what the class is wearing that day.

FOLLOW-UP

⊗ Have groups share and compare their fraction analyses.

⊗ Have any five students stand where all can see them. Ask a volunteer to write one fraction that describes something about the clothing those students are wearing. Then challenge other students to identify the feature of the students' dress that the fraction describes. Repeat with other volunteers.

CLASS FRACTION WALL

How can you use fractions to describe your class from one day to the next?

GOAL: **Students create an ongoing bulletin board display that presents a profile of the class. The display describes classmates' interests and preferences.**

MATERIALS: *posterboard, markers*

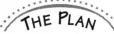 THE PLAN

1. Explain to students that fractions can be used to create a profile of the class. Use the example of favorite foods. Generate a list of five foods the class enjoys. Record how many students prefer each of the foods. Then guide them to use fractions to express those preferences. For example, if 5 students name pizza as their favorite food, and there are 21 students in the class, the fraction $\frac{5}{21}$ represents that information: The numerator is the number of students who expressed a particular preference; the denominator is the total number of students in the class.

2. As needed, review how to use tally marks to record responses in a survey.

3. Divide the class into groups, assigning each a different topic generated by whole-class discussion. Examples might include favorite sports, ice cream flavors, TV characters, ways to spend free time, authors, or seasons.

4. Provide each group with posterboard and markers. Have each group conduct a survey of their class in which students choose the item they most prefer from a list of choices the group generates and presents. Tasks include developing the list of choices, polling all students, tallying responses, making a chart based on the results, and describing the process.

5. Invite groups to post their charts on the bulletin board. Use the data to formulate questions that students can answer with a fraction.

 FOLLOW-UP

⊗ Repeat and revise this class profile activity as an ongoing, ever-changing investigation. Change topics, student groupings, and ways to use fractions to interpret results. For example, students can compare results to one-half, to one another, and to results from earlier surveys or from those done in other classes.

⊗ Challenge students to create computer-generated charts or tables to represent the data they have gathered. Compare and contrast different types of graphic organizers.

PICTURE THESE FRACTIONS

How can fractions help you describe what you see in pictures?

GOAL: Students create a display of pictures or photos of groups. They use fractions to describe what the pictures show.

MATERIALS: old magazines or newspapers; loose photos of groups of people, animals, or various items; scissors; tacks or staples

THE PLAN

1. Create a bulletin board display consisting of group photos. Post pictures of groups of people or animals, or of discrete objects that can be counted, such as bunches of assorted flowers, bowls of fruit, books on shelves or scattered on tabletops, or various toys or gadgets in store window displays.

2. Post a set of questions about the pictures that students can answer using fractions. For example, for a photo of fans at a ball game you might ask, "What fraction of the fans are children? What fraction of the fans are wearing sunglasses? What fraction are seated?"

3. Keep updating the bulletin board display. Change the photos frequently, along with the questions that accompany them. Involve students in the creation of the display by inviting them to bring in their own photos or to find some in the periodicals you provide. Have them post fraction questions with their group photos for classmates to answer.

FOLLOW-UP

⊗ Play a guessing game to extend the lesson. Form small groups. One member secretly chooses one of the photos on display and describes some feature of that photo using fractions. For example, "Half the boys in one photo are wearing baseball caps. Which photo am I describing?" The other students use the clue to identify the photo. Students can swap roles.

⊗ If you have access to digital photography, use the editing features to generate variations on class photos to reinforce the concept of fractions of a set.

Mixing Snack Mix

Try a hands-on approach to fractions of a set.

Goal: Students use counters to create "snack mixes" that satisfy fraction rules.

Materials: 18 counters (or other small objects) in each of 3 colors, student page 29

Teaching Tips

- Provide each student or pair with counters or other small objects with which to work out these fraction challenges. If you wish to provide actual raisins, nuts, and chocolate chips, be sure that students wash their hands first and that they don't eat the spoils until the end!

Fraction Message

Identify fractions of a set to answer a geography question.

Goal: Students identify fractions of a set—in this case, particular letters within a word.

Materials: student page 30

Teaching Tips

- Duplicate and distribute copies of page 30 to each student or pair. Explain that students must correctly identify the fractional part of each word and write the correct letter in the space to its right. When they have found all the letters, they can copy them in order at the bottom of the page to answer the question about the Potato Museum.

The Language of Fractions

Complete sentences using fraction words.

Goal: Students apply their understanding of fraction concepts in order to complete sentences.

Materials: student page 31

Teaching Tips

- Duplicate and distribute copies of page 31 to each student. Explain that each word in the box will be used only once. Invite students to develop variations of this page for classmates.

MIXING SNACK MIX

Try to make a snack mix according to fraction rules!

You will need 18 counters of one color, 18 of a second color, and 18 of a third color. Put them in groups (or cups) on your desk.

⊗ Let one set of counters stand for raisins.

⊗ Let the second set stand for nuts.

⊗ Let the third set stand for chocolate chips.

Use counters to make different snack mixes according to the rules given. Record how many of each "food" is in your mix.

1 Make an 18-piece snack mix with raisins and nuts. Use half as many nuts as raisins.   _____

2 Make a 24-piece snack mix with nuts and chocolate chips. Use one-third as many chips as nuts.   _____

3 Make a 36-piece snack mix with raisins, nuts, and chocolate chips. Use the same number of each food.   _____

4 You want to make a 48-piece snack mix using all three snacks. You want to use all the nuts and half as many raisins as nuts. Are there enough chocolate chips for this mix? Explain.

_____

5 Jorge made a snack mix with 16 chocolate chips, 12 raisins, and 4 nuts. Use fractions in all the ways you can to describe Jorge's mix.

_____

6 Create your own sets of rules for making a snack mix. Write down your rules. Give them to classmates to use.

PREPARE TO SHARE ....

What was tricky about this task? Explain.

FRACTION MESSAGE

Peel yourself away from the other attractions in this city so you can enjoy its Potato Museum. What is the name of this city that takes its spuds so seriously?

To find out, identify the fraction of each word. Write the letters in order on the lines below. The first one has been done for you.

1. the first $\frac{1}{3}$ of ANT ___A___

2. the last $\frac{1}{4}$ of GIRL _____

3. the second $\frac{1}{5}$ of ABOUT _____

4. the second $\frac{1}{4}$ of QUIT _____

5. the middle $\frac{1}{5}$ of PIQUE _____

6. the first half of UP _____

7. the middle $\frac{1}{5}$ of BREAK _____

8. the second $\frac{1}{5}$ of PROVE _____

9. the second fourth of AQUA _____

10. the fourth fifth of ABOUT _____

11. the second fifth of READY _____

12. the second half of IN _____

13. the final fourth of PINE _____

14. the last $\frac{1}{4}$ of FLEW _____

15. the middle third of AMP _____

16. the last sixth of PUDDLE _____

17. the last third of TAX _____

18. the second fourth of DIET _____

19. the first $\frac{1}{6}$ of CHROME _____

20. the third fifth of GROOM _____

PREPARE TO SHARE

The Potato Museum is in_____.

Mega-Fun Fractions ✳ Scholastic Professional Books

THE LANGUAGE OF FRACTIONS

Here are some words about fractions.

Use the word from the box that best completes each sentence.

half	quarter	eighth	denominator
third	fifth	tenth	equivalent
fourth	sixth	numerator	improper

1. The number above the bar in a fraction is the one you say first. This number is called the _____.

2. One of five equal parts is known as a _____.

3. One of six equal parts of something is a _____.

4. One _____ is the same as one-fourth of a dollar.

5. A fraction is said to be _____ when its numerator is greater than its denominator.

6. Either of two equal parts of a figure is _____.

7. Two fractions are _____ if they name the same amount.

8. In a fraction, the _____ tells the total number of parts.

9. One-_____ is less than one-sixth but greater than one-ninth.

10. A year is one-_____ of a decade.

11. One-_____ is greater than one-fourth but less then one-half.

12. Each side of a square is one-_____ of the distance around the square.

WRITE ABOUT IT

Look around the room for things you could describe using fraction words. Then write five sentences about what you see. In each sentence, use a different word from the box. Underline the word you chose. Example: The window is open about one-<u>fifth</u> of the way.

Mega-Fun Fractions ✸ Scholastic Professional Books

FRACTION DICTATION

Can you write fractions and mixed numbers that you hear spoken aloud?

GOAL: Students write the numerical representation of a fraction or mixed number when they hear it spoken.

MATERIALS: none

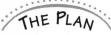

THE PLAN

1. Begin by discussing with students the importance of knowing the correct way to write a fraction or mixed number. Reiterate that most fraction word names end with *-ths*, with the exception of *halves* and *thirds*.

2. Read aloud sentences containing fractions or mixed numbers. Alternatively, have students form pairs or small groups and give sentences to a member of the group to read aloud to the others.

3. When students hear the sentences, they write in numerical form the fraction or mixed number they hear. Try several practice sentences so that students grasp the task.

4. Use these sentences or others like them:

   ⊗ More than **one-half** of the world's population lives in Asia.

   ⊗ **Two-sevenths** of the days of the week begin with *T*.

   ⊗ Each week in February is **one-fourth** of the month, except during leap years.

   ⊗ The value of a quarter is **two and one-half** times the value of a dime.

   ⊗ A basketball season can last about eight months, or **two-thirds** of a year.

   ⊗ In the American Football Conference, **two and one-fifth** times as many teams do not make the playoffs as do make them.

   ⊗ The Thanksgiving Tower in Dallas has **five-sixths** as many floors as the city's Bank One Center does.

   ⊗ **One-third** of a healthy diet comes from fruits and vegetables.

   ⊗ I ran **seven-eighths** of a mile in five minutes.

   ⊗ **Eleven-twelfths** of the months have at least 30 days.

   ⊗ Each day, on average, the *Wall Street Journal* sells about **four and one-fourth** as many newspapers as the *Newark Star-Ledger* does.

   ⊗ **Ninety-eight-one-hundredths** of all American homes have at least one television.

FOLLOW-UP

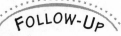

⊗ Invite students to make up their own sentences containing fractions. Have them say these aloud to a partner, who must then write the fraction correctly.

Fractions and Egg Cartons

How can an empty egg carton model the idea of equivalent fractions?

GOAL: Students use egg cartons to get hands-on experience with equivalent fractions.

MATERIALS: empty egg cartons, beans or counters, pipe cleaners or yarn (optional)

The Plan

1. Distribute empty egg cartons to pairs of students. Point out that each carton represents one whole. Elicit from students what fraction each cup in the carton represents. (*$\frac{1}{12}$*)

2. Ask students to use the egg carton to model different fractions. For example, have them place beans, one per cup, to show $\frac{1}{12}$, $\frac{5}{12}$, and $\frac{11}{12}$. Ask them to tell how many beans they need to show each fraction they make. Ask them to tell how many beans they would need to model the fraction $\frac{12}{12}$. (*12*)

3. Have pairs of students place beans, one per cup, in their carton to model a carton half-filled. They may place one bean in each cup in any order to show one-half. Ask, "Count your beans. How many beans have you used to show one-half?" (*6*) Then ask, "Does it matter where you place the beans?" (*no*) "How many twelfths make one-half?" (*$\frac{6}{12}$*)

4. In the same manner that you guided students to see that $\frac{6}{12}$ is equivalent to $\frac{1}{2}$, guide pairs to place beans in their cartons to model the equivalence of $\frac{4}{12}$ and $\frac{1}{3}$. Then repeat for $\frac{3}{12}$ and $\frac{1}{4}$, and for $\frac{2}{12}$ and $\frac{1}{6}$. Each time, have students explain what they did to model the fraction. Be sure pairs empty their cartons before modeling each new equivalence. If appropriate, guide students to establish the number of cups in three equal groups, four equal groups, and six equal groups. This will help them model each fraction equivalence.

5. Have students use the beans to fill two-thirds of the carton. Before they begin, invite them to speculate about how many beans this will take and ask them to give their reasoning. Then ask, "How many twelfths make two-thirds? How do you know?" (*$\frac{8}{12}$*)

6. Repeat this procedure for the fractions $\frac{3}{4}$ and $\frac{5}{6}$.

7. Ask students to summarize what they have learned about fractions in this activity. Challenge them to think of another hands-on way to explore the same idea.

Follow-Up

- Instead of using beans or counters, have students use pipe cleaners or yarn to show fractions of the egg carton (as a region). For example, to show half, they can wrap the pipe cleaner around the middle of the carton, separating it into two sets of six cups.
- To explore mixed numbers, such as $1\frac{1}{4}$, pairs can use two or more egg cartons to conduct similar hands-on modeling.

BADGE BUDDIES

How do you know which fraction badge is a match for yours?

GOAL: Students play a game in which they find equivalent fractions.

MATERIALS: large index cards, pieces of oaktag, or construction paper; string or yarn; staples; shopping bags

THE PLAN

1. Make sets of fraction badges using the index cards, oaktag, or construction paper. Each set should consist of at least three equivalent fractions, such as $\frac{1}{3}$, $\frac{2}{6}$, and $\frac{3}{9}$. Prepare enough sets to allow one badge per student. Staple string or yarn to each badge so that students can wear them around their necks, hanging loosely across the chest.

2. Mix up the badges and place them in a shopping bag. Then have students reach in, take a badge, and put it on.

3. Give students three minutes to find their badge buddies—all other classmates whose badges display a fraction equivalent to the one the student is wearing.

4. Invite students to call out to each other during the buddy search, as they look for specific fractions they know belong in their group. Groups stand or sit together when they are complete.

5. Have students return their badges to the badge bag after a round. Then play again.

FOLLOW-UP

⊗ Play the matching game to reinforce the concept of fractions in simplest form. When you do this, make sure that one fraction in each set is expressed in lowest terms. When students have found their badge buddies, ask them to indicate the student who is wearing the badge with the fraction in simplest form.

⊗ Include fractional numbers greater than one. Make sets of equivalent mixed numbers and improper fractions, such as $\frac{13}{4}$, $\frac{14}{8}$, and $\frac{19}{12}$, and include them in the badge bag.

EQUIVALENT FRACTION CONCENTRATION

Can you spot equivalent fractions when you see them?
Can you recall where they are on a grid?

GOAL: **Students play a familiar game in which they seek to match equivalent fractions.**

MATERIALS: *index cards*

THE PLAN

1. Tell students that they are going to play Equivalent Fraction Concentration. Have them help you make the playing decks, each of which will have 20 pairs of equivalent fraction cards. Divide the class into groups of four and have each group make a deck of cards. Each student should write a different fraction on each of five cards and then write the equivalent of each one on another five cards.

2. Instruct students to shuffle their decks of 40 index cards and then turn them all facedown on the table in an 8 x 5 array.

3. Present these rules:

   ⊗ Establish an order of play.

   ⊗ Taking turns, each player flips two cards faceup. If the cards display two equivalent fractions, it is considered a match. The player removes those cards from the table and takes another turn.

   ⊗ If the cards do not match, the player turns them facedown in their original positions and the next player takes a turn.

   ⊗ The winner is the player who has collected the greatest number of cards once all matches have been made.

4. Have groups play several times, making sure to shuffle the decks well after each game.

FOLLOW-UP

⊗ Have students play with decks made by other groups.

⊗ Increase or decrease the number of sets in each deck.

⊗ Extend the game by including decimals and/or money amounts equivalent to fractions in the decks—for example, sets such as 0.5 and $\frac{1}{2}$, $.80 and $\frac{4}{5}$.

⊗ Extend the game by including mixed numbers and their equivalent improper fractions in the decks.

FRACTIONS BINGO—TIMES TWO

How well can you match equivalent fractions or round mixed numbers?

GOAL: Students practice matching equivalent fractions, or mixed numbers and improper fractions, using a Bingo format.

MATERIALS: student pages 37 and 38, counters, 1–6 number cubes (2 per group, p. 85), 1–8 spinner (1 per group, p. 90)

THE PLAN   EQUIVALENT FRACTIONS BINGO

(1) Have students form groups of four, and review the basic rules for Bingo. Tell groups that one member in the group will be the caller. Then duplicate and distribute the Equivalent Fractions Bingo student page. Provide each group with two 1–6 number cubes and a stack of counters with which to cover squares on their cards.

(2) Review the directions on the student page. Guide students as they fill in their grids. Then present the rules for playing:

⊗ The caller rolls the two number cubes and calls out the fraction created. The smaller number is the numerator and the greater one is the denominator.

⊗ Players scan their cards for a fraction equivalent to the fraction called. If they find an equivalent fraction, they cover its space with a counter.

⊗ The first player to cover a row, a column, or a diagonal on his or her card wins. The winner becomes the caller for the next round of Bingo.

THE PLAN   ROUNDING MIXED NUMBERS BINGO

(1) Duplicate and distribute the Rounding Mixed Numbers Bingo student page. For this game, each group needs a spinner as well as two number cubes. Then present the following game rules for rounding mixed numbers: Compare the fraction part to $\frac{1}{2}$. If it is less than $\frac{1}{2}$, round down. If it is greater than or equal to $\frac{1}{2}$, round up.

(2) Present the rules for this version of the game:

⊗ The caller spins the spinner and rolls the two number cubes. The number indicated on the spinner is the whole-number part of the mixed number. The smaller number rolled is the numerator of the fraction part, and the greater number rolled is the denominator. For example, if the caller spins a 3 and rolls a 1 and a 5, the mixed number would be $3\frac{1}{5}$. Have the caller write down the number.

⊗ Players round the mixed number to the nearest whole number, and then scan their cards to find that number. If they find the number, they cover its space.

⊗ The first player to cover a row, a column, or a diagonal on his or her card wins. The winner becomes the caller for the next round.

EQUIVALENT FRACTIONS BINGO

Choose 16 of the fractions from the box below. Write one fraction in each square of your Bingo card. You may use any fraction one or two times, but not more than that. Write the fractions in any squares you wish.

| $\frac{2}{12}$ | $\frac{2}{10}$ | $\frac{3}{15}$ | $\frac{2}{8}$ | $\frac{3}{12}$ | $\frac{3}{9}$ | $\frac{5}{15}$ | $\frac{5}{10}$ | $\frac{7}{14}$ | $\frac{4}{10}$ | $\frac{6}{15}$ |
| $\frac{6}{10}$ | $\frac{9}{15}$ | $\frac{6}{9}$ | $\frac{8}{12}$ | $\frac{6}{8}$ | $\frac{9}{12}$ | $\frac{8}{10}$ | $\frac{12}{15}$ | $\frac{10}{12}$ | $\frac{15}{18}$ | |

My Fractions Bingo Card

PREPARE TO SHARE ....

How would you vary the way you fill your card for future games? Explain.

ROUNDING MIXED NUMBERS BINGO

Write one number from 1 to 9 in each square of your Bingo card. You may use each number one, two, or three times, but not more than that. You may place the numbers in any squares you wish.

My Fractions Bingo Card

PREPARE TO SHARE

How might you fill out your card differently for future games? Explain.

Mega-Fun Fractions ⊗ Scholastic Professional Books

FRACTION POEMS

How can these poems help you remember fraction rules?

GOAL: Students read, recite, and present poems to reinforce fraction concepts.

MATERIALS: student page 40, tape recorder or CD burner (optional)

THE PLAN

1. Duplicate and distribute copies of page 40. Invite students to read the poems silently to get the general idea of what each one means.

2. Discuss with students the concept or idea behind each poem. For example, the poem entitled "Half" addresses the mistaken idea that a person can get a "bigger half." By definition, *half* means either of two equal parts, so mathematically, there cannot be a bigger half. Work through a fraction exercise with the related poem on hand. Guide students to see how the words of the poem relate to the steps they follow.

3. Invite students to think of ways to present the poems. For example, they might chant them, set them to music, recite them over background rhythms, or arrange them for choral reading. Poems can be read or memorized, as you see fit.

FOLLOW-UP

⊗ Challenge students to create their own poems that address fraction concepts. Or invite them to add verses to the poems presented here.

⊗ Extend this activity by producing a class fraction tape (or CD) that presents these poems along with those students have written.

FRACTION POEMS

Read these fraction poems. Think about what each one means.

Be Fair!

Be fair to your friend, the friendly fraction,
Reduce both terms fairly when you try.
Be fair to numerators and denominators,
'Cause you'd never want to make a fraction cry!

Be fair to both terms when you reduce them,
Reduce them equally to simplify.
Be fair to numerators and denominators,
'Cause you'd never want to make a fraction cry!

Half

You may think it's fun to laugh
When you get the bigger half.
But, my friend, the laugh's on you:
Half means equal parts in two.

Lowest Common Denominator

Lowest common denominator,
What does that mean to me?
Lowest common denominator,
It's the good old LCD.

Wanna add or subtract two
  fractions?
It's easy if you play the game.
To add or subtract two fractions,
The denominators must be the
  same.

So how do you get two fractions
To be in synch for you?
Find the lowest common
  denominator,
And make the switcheroo.

GCF

Simplest form, simplest form,
How on earth will you transform
A fraction with greater terms to be
Of equivalent value, and easy to see?

GCF, GCF,
Greatest common factor, that's what's left.
Put fractions on a diet with a GCF,
Reduce those terms with a GCF.

Mega-Fun Fractions ⊗ Scholastic Professional Books

SHARING FRACTION PIE

How can you use clay pies to compare fractions?

GOAL: Students create, cut, and serve clay "pies" to explore comparing fractions.

MATERIALS: modeling clay, pencils (or dowels) for rolling pins, same-sized plastic lids for pie plates (from yogurt cups, margarine tubs, coffee cans, and so on), plastic knives, paper plates

THE PLAN

1. Divide the class into pairs or small groups. Provide each group with modeling clay, pencils or dowels, same-sized plastic lids, plastic knives, and paper plates. Tell students that they are going to make clay "pies" to cut into fractional parts and then compare the sizes of the parts.

2. Have children roll out the clay "pie dough" to fill the plastic lid. Then have them use the plastic knives to cut the pie in half and serve $\frac{1}{2}$ on a paper plate. Ask another member of the group to cut his or her pie into fourths and serve $\frac{1}{4}$ on the same plate. Ask: "Which is greater, $\frac{1}{2}$ or $\frac{1}{4}$? How do you know?" Record the answer on the chalkboard, using the mathematical symbol for *greater than* (>). ($\frac{1}{2} > \frac{1}{4}$)

3. Repeat this process with other fractional parts students can cut from their clay pies. Remind them to reroll the dough to begin with a whole pie each time. Generate a list of fraction comparison statements, such as $\frac{1}{3} > \frac{1}{6}$, $\frac{2}{3} > \frac{1}{2}$, $\frac{3}{4} > \frac{3}{5}$, that derive from students' explorations.

FOLLOW-UP

⊗ Extend by having students make, slice, and serve clay pie slices to form fraction comparison statements that use the *less than* symbol (<). For instance, $\frac{1}{2} < \frac{3}{4}$.

⊗ Challenge students to compare halves of pies of different sizes to reinforce the concept that halves of same-sized regions are always equal, but halves of different-sized regions may well vary in size from one another.

FRACTION FILL 'EM UP

Why does a fraction with a larger denominator represent a smaller part of the whole if the numerators are the same?

GOAL: Students do a hands-on exploration to compare fractions with unlike denominators.

MATERIALS: sets of four glass jars, masking tape, markers, pitchers, water, food coloring, paper towels

THE PLAN

1. Gather sets of four same-size glass jars, like mayonnaise, baby food, or peanut butter jars, for each group of 3–4 students. For each set, measure and mark off one jar each with masking tape to show halves, thirds, fourths, and fifths.

2. Divide the class into groups. Provide each group with a set of jars, a pitcher filled with colored water, and paper towels in case of spills.

3. Instruct groups to pour from the pitcher to fill each jar as follows:

   ⊕ Fill the jar marked in halves to $\frac{1}{2}$.　　⊕ Fill the jar marked in fourths to $\frac{1}{4}$.

   ⊕ Fill the jar marked in thirds to $\frac{1}{3}$.　　⊕ Fill the jar marked in fifths to $\frac{1}{5}$.

4. Have students compare the amount of liquid each jar holds, two jars at a time: Compare the halves jar with the thirds jar, the thirds jar with the fourths jar, and the fourths jar with the fifths jar. Ask students to explain what they notice.

5. Have students compare all four jars by placing them in order from least to most full. Again, ask them to describe what they notice.

6. Have students empty the water from the jars into the pitcher. Now have them refill the jars. This time, they should fill $\frac{2}{2}$ of the halves jar, $\frac{2}{3}$ of the thirds jar, $\frac{2}{4}$ of the fourths jar, and $\frac{2}{5}$ of the fifths jar. Before they pour, ask them to predict which jar will have the most liquid in it and which will have the least. Have students compare jars two at a time and order them from least to most filled-up.

7. Finally, ask students to summarize what they have observed about the sizes of fractions when the numerators (parts being considered) are the same but the denominators (total number of parts) differ.

FOLLOW-UP

⊕ Extend by having students use the jars to compare various fractions. For example, they can fill the jars to see that $\frac{2}{5} > \frac{1}{3}$ but $< \frac{1}{2}$ or that $\frac{3}{4} > \frac{1}{2}$ but $< \frac{4}{5}$.

⊕ Extend by having students use calibrated measuring cups to perform similar comparisons using sand, water, or rice.

PROVE IT!

How can you prove that one fraction is greater than another?

GOAL: Students use models, drawings, oral arguments, and demonstrations to "present a case" for the inequality (or equality) of two given fractions.

MATERIALS: counters, containers, fraction strips (p. 84), rulers, pattern blocks (p. 86), paper and pencil, grid paper (p. 88 and 89), index cards

THE PLAN

1. Explain to students that in legal trials, lawyers make opening statements, present evidence, and then offer logical conclusions in their closing arguments. Tell students that they will pretend to be math lawyers.

2. Divide the class into groups of 4–6 lawyers, each headed by lead counsel. Tell groups that their assignment is to use the elements of a trial and all the resources available to them in the classroom to present a full case before a jury of peers (their classmates) to prove the inequality or equality of a pair of fractions.

3. Provide each legal team with an index card on which you have written a fraction statement to prove. Choose from the following statements, or provide others like them: $\frac{1}{3} > \frac{1}{6}$, $\frac{2}{9} < \frac{5}{9}$, $\frac{3}{4} < \frac{11}{12}$, and $\frac{1}{4} = \frac{2}{8}$. Challenge groups to work cooperatively, employing a variety of tools and strategies, to prepare a case to present to the class. Guide them to make their cases as clear, logical, and thorough as possible.

4. Allow groups time to develop their case and to plan strong opening and closing statements. Encourage groups to plan carefully the order of evidence to be presented and to involve all members in the presentation. Then have them present their case to the class. You may wish to videotape group presentations for future use as teaching tools.

FOLLOW-UP

⊗ Allow time for students to comment on the effectiveness of groups' cases. Invite them to argue why the presentations were or were not convincing and to comment on what elements of the cases were most (and least) effective.

⊗ Extend group assignments to include inequality statements involving mixed numbers and improper fractions or fractions and decimals.

FRACTION WAR

How can playing a game reinforce your ability to compare fractions?

GOAL: Students play a card game in which they create and then compare fractions.

MATERIALS: sets of 24 index cards (two sets numbered 1 to 12)

THE PLAN

1. This is the first of three games in which students compare or order fractions. Play is based on the classic card game War, with which your students are no doubt familiar. Players do not manipulate their cards; they simply compare the two fractions formed by overturned cards.

2. Have students form pairs. Provide each pair with a deck of 24 cards. To play, students shuffle their deck of cards and deal them so that each player gets 12 cards facedown in a stack.

3. To begin, each player turns over the top two cards from his or her stack. Players arrange the two cards to form a fraction, with the greater number as the denominator. Players then compare the fractions that have been formed. The player with the greater fraction wins. That player then takes all four cards and places them at the bottom of his or her stack.

4. If the two fractions are equivalent, then neither player wins the cards. Those cards remain on the table as each player turns over two more cards and again builds a fraction. The player with the greater fraction now collects all eight cards on the table for his or her stack. Just like in the game of War, the player who collects all the cards wins.

5. At the conclusion of this activity, invite students to give their opinions of the rules of play. Invite suggestions about how to improve the game.

FOLLOW-UP

⊗ Have students make up their own decks for Fraction War. Invite them to use more or fewer than 24 cards and any whole numbers they choose.

⊗ Vary the rules so that the player with the lesser fraction wins. Or vary the game to include three players; the winner can be the player with the greatest (or least) fractional number. If there is a tie between two players, only those two turn over another pair of cards, make new fractions, and compare again.

⊗ For a more challenging game of Fraction War, create a deck of cards that consists of fractions and decimals. There should be an equal number of each form of number, and the values should be equivalent. For example, a deck with $\frac{1}{2}$, $\frac{3}{5}$, and $\frac{7}{8}$ would also have 0.5, 0.6, and 0.875.

FRACTION ROLLERS, PART 1

Given two digits, can you make the greatest fraction? The least fraction?

GOAL: Students play a game in which they create and compare large and small fractions.

MATERIALS: 1–6 number cubes (p. 85), index cards

THE PLAN

1. Have students form pairs. Provide each pair with two number cubes. Tell students that they will play a game in which they create fractions with a roll of the number cubes.

2. Present the rules for play:

   ⊗ Each player rolls the number cubes to get two digits.

   ⊗ The player uses those digits to write the greatest possible fraction the digits can make. (This will always be a whole or mixed number.)

   ⊗ The player with the greater number wins a point; no points are given for a tie.

   ⊗ The first player to reach 10 points wins.

3. Have pairs play several times to provide adequate practice and to sharpen their understanding of the strategy of the game.

4. Vary the rules so that players try to form the *least* possible fraction using the two digits rolled. In this variation, the winner each time is the player with the smaller fraction. Or vary the game to include three players; in this case, the winner can be the player with the greatest (or least) fractional number or the one with the number in the middle.

5. For a more challenging game, have players use three number cubes to build their numbers in any fashion they wish. For instance, a roll of 1, 3, and 6 can be interpreted not only as a mixed number such as $1\frac{3}{6}$, $3\frac{1}{6}$, or $6\frac{1}{3}$, but also as $\frac{1}{36}$, $\frac{1}{63}$, $\frac{6}{13}$, and so on.

6. For whichever version of the game your students play, have them summarize the components of a winning strategy.

FOLLOW-UP

⊗ Have students make up their own comparing-fractions game in which they use cards, number cubes, spinners, or other manipulatives. Guide them to think through their games and test them before writing down the rules on index cards. Then have students teach their games to classmates. Invite them to demonstrate how to play.

⊗ Invite students to contribute their ideas to a class book of fraction games.

THAT'S AN ORDER!

How well can you predict whether a fraction will be greater than, less than, or in between the value of two other fractions?

GOAL: Students play a fraction card game that helps them gain a solid understanding of how to compare and order fractions.

MATERIALS: sets of 48 index cards

THE PLAN

1. Prepare sets of 48 fraction cards, with halves, thirds, fourths, fifths, sixths, eighths, tenths, and twelfths in each set. The numerators can be any numbers that form fractions less than or equal to 1. Include $\frac{2}{2}$, $\frac{3}{3}$, $\frac{4}{4}$, and so on.

2. Tell students that they will play a game in which they must put fractions in size order. Have students form groups of four. Provide each group with a set of fraction cards. Tell them that in each game, three players play and the fourth is the dealer and judge. (These roles can shift.)

3. Present the rules for play:

   ✪ The dealer shuffles the deck, then deals each player two fraction cards faceup and a third card facedown.

   ✪ Players examine their faceup fraction cards and order them from lesser to greater. Then they decide whether the third card is likely to be less than both cards, greater than both cards, or greater than one card but less than the other. Each player states his or her prediction by saying "Greater," "Less," or "In between."

   ✪ Then the dealer turns the third card faceup. Players each get 1 point if their prediction was correct.

   ✪ The first player to earn 7 points wins.

4. Allow students to work with paper and pencil if they wish. Groups should play enough games so that each member gets a chance to play and to deal.

FOLLOW-UP

✪ Encourage students to articulate a winning strategy for this game. Invite them to share pointers with classmates.

✪ Simplify the game to omit the prediction students must make, but leave the task of ordering fractions. In this variation, each player gets a point if and only if the third card has a value in between the other two.

✪ Extend the game by having the dealer deal all three cards at once. Players get 1 point if they correctly order their cards from least to greatest. Try this with four-card deals, too.

FRACTIONS OF A DAY

How can fractions help you make a circle graph of how you spend a day?

GOAL: Students fill in a chart to show how they spend their day. Then they use the data in the chart to create a circle graph. They use fractions to interpret what the charts show.

MATERIALS: student pages 48–49, crayons

THE PLAN

1. Brainstorm with students a list of the main kinds of activities kids do every school day, almost every day, and on some days but not others. Then tell students that in this activity, they will estimate how much time they themselves actually spend on these activities, record that information in a table, and then display it on a circle graph.

2. Explain that a circle graph is a useful tool for showing how an amount of data is divided.

3. Duplicate and distribute the Fractions of a Day student pages and hand out crayons. First, go over how to complete the table on page 48. Guide students to begin by recording how much time they spend doing everyday activities of sleeping, attending school, and eating meals. Also talk about the idea of establishing average daily times for activities that vary in length from day to day. Remind students that the total time for all daily activities must add up to 24 hours.

4. Help students understand how to complete the circle graph on page 49. Guide them to notice the 24 divisions—one per hour of a day. Direct students to use a different color for each activity and to label it—either within its part of the circle or outside it, indicated with a line or arrow.

5. Have students use fractions to summarize the information on their graphs. For example, a student who sleeps 8 hours a day sleeps for $\frac{8}{24}$ of the day. If appropriate for your class, have students express their fractions in simplest form.

6. When students finish, help them post their graphs and summaries on a bulletin board. Invite volunteers to analyze the data on the graphs and to discuss similarities and differences among them.

FOLLOW-UP

⊗ Have students collaborate on a class graph of the same data.

⊗ Invite students to make tables and graphs that show how they typically spend their time in school or on a weekend day. Help them generate a circle chart with the appropriate number of sectors for the graph they wish to display.

FRACTIONS OF A DAY, PART 1

Fill in the chart with the activities you do on a typical school day. Assign each activity a different color. The total time for all activities must add up to 24 hours.

Activity	Time Spent (to nearest hour)	Color

PREPARE TO SHARE

In what ways would this table look different if you filled it in for any given Saturday? Explain.

FRACTIONS OF A DAY, PART 2

Use the information from the chart on page 48 to make a circle graph. Color a section of the circle graph to show how much time you spend doing each activity. Fill in each section with the color you chose for that activity. Label each section of the graph with the name of the activity and a fraction that describes the amount of time.

How I Spend My Day

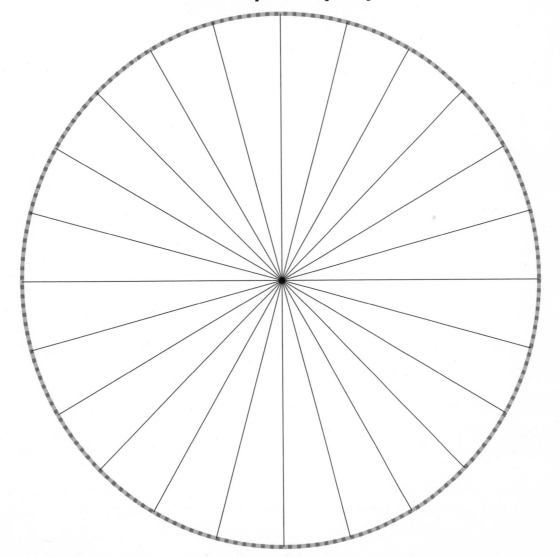

∴PREPARE TO SHARE∴ ··

In what ways would this graph look different if you based it on how you might spend a day when you're sick? Explain on the back of this page.

TIME FOR FRACTIONS

How can you use fractions to express lengths of time?

GOAL: Students add amounts of time to reinforce the relationship between time and fractions (parts of an hour).

MATERIALS: sets of 20 index cards , play clocks or analog watches

THE PLAN

1. Make sets of 20 index cards (one set per group) for the following times:

$\frac{1}{6}$ hour	$\frac{1}{2}$ hour	10 minutes	40 minutes	sixth of an hour
$\frac{1}{5}$ hour	$\frac{2}{3}$ hour	15 minutes	45 minutes	quarter of an hour
$\frac{1}{4}$ hour	$\frac{3}{4}$ hour	20 minutes	48 minutes	thirty minutes
$\frac{1}{3}$ hour	$\frac{5}{6}$ hour	30 minutes	50 minutes	half an hour

2. Have students form groups of 3–4. Tell groups that they will play a game in which they record and use amounts of time, expressed in minutes or in fractions of an hour. Provide a play clock (such as a paper-plate clock) or analog watch that students can manipulate to show accumulated time.

3. To play, students shuffle their cards and place them facedown in a stack. They start their group's clock at 9:00 A.M. One player turns over the top card, reads aloud the amount of time it shows, and moves the hands of the clock or watch ahead to show the new time. Other group members verify that the clock hands are correctly positioned. The used card is returned to the bottom of the stack.

4. Play continues in turn, with each player drawing a card, reading the amount of time it shows, then moving the clock ahead to show the amount of time that has passed.

5. The player who is the one to draw a card that moves the clock to 3:00 P.M. or beyond wins.

FOLLOW-UP

✸ Conclude the activity by talking about the game. Discuss what was easiest and hardest about it. Discuss ways to change or improve the game. Are there cards that students would add to or remove from the set? Which ones?

✸ For an additional challenge, add cards to the sets—for example, more fifths as well as tenths and twelfths.

✸ Challenge students to compute backward from, for example, 6:00 P.M. to noon.

✸ Help students create a similar game using fractions of a year (as well as time given in days, weeks, and months), in which they move on a calendar from January to December.

COINING FRACTIONS

How can you express money amounts in fractional terms?

GOAL: Students use play money to represent parts of a dollar and to build an understanding of renaming improper fractions as mixed numbers.

MATERIALS: sets of paper coins (page 91) or play coins (5 pennies, 5 nickels, 5 dimes, and 5 quarters per set)

THE PLAN

1. Group students into pairs. Provide each pair with a set of coins.

2. Discuss the value of each coin as a fraction of a dollar. Guide students by asking questions, such as "What fraction of a dollar is a dime?" or "What fraction of a dollar is a quarter?" Then ask, "What fraction of a dollar is a quarter plus a dime?" or "What combination of the coins in your set makes the fraction $\frac{1}{5}$ of a dollar?" or "Which three coins add to make the fraction $\frac{27}{100}$ of a dollar?"

3. Give partners a few minutes to generate questions like these for each other. Guide them to manipulate the play coins to answer the questions they formulate.

4. Guide students to understand that a money amount over $1.00 can be interpreted as a mixed number. For example, $1.20 = 1 whole dollar and $\frac{20}{100}$ of a dollar, or $1\frac{20}{100}$, or $1\frac{1}{5}$.

5. Have one partner gather some of the coins and place them on the desktop, grouped by type. Ask that student to give the sum, in dollars and cents, of the coins chosen. The other partner then determines what fraction of a dollar is represented by each type of coin picked. For example, 2 quarters can be expressed as $\frac{2}{4}$ (or $\frac{1}{2}$) of a dollar. Then the same partner writes each fraction as an addend in an addition problem. For example, if the first student were to select 2 quarters, 3 dimes, 1 nickel, and 2 pennies (for a total of $.87), the second student would write: $\frac{2}{4} + \frac{3}{10} + \frac{1}{20} + \frac{2}{100}$. If appropriate, have students express each fraction in simplest form.

6. Have partners take turns collecting and grouping the coins and writing an addition problem for each set.

FOLLOW-UP

⊗ Challenge the second partners to find the sum of the coins by adding the fractions. For example, for the problem given above, a student might write:

$$\frac{50}{100} + \frac{30}{100} + \frac{5}{100} + \frac{2}{100} = ? \left(\frac{87}{100} \right)$$

⊗ Have students each take two discrete bunches of coins. Have them use fractions to represent the two money amounts and then find their difference.

⊗ Duplicate and distribute the following page, an activity that challenges students to further explore the relationship between fractions and money amounts.

FUNNY MONEY

You will need the coins shown in the above picture. Use them to solve these problems.

1. How many coins are there in the whole set? _____

2. One-half of the coins in this set have a value of 41¢.

   How many coins is this? _____

   Which coins are they? _____

3. One-quarter of the coins add up to 6¢. How many coins is this? _____

   Which coins are they? _____

4. Three-quarters of the coins make 63¢. How many coins is this? _____

   Which coins are they? _____

5. What is the least value that $\frac{1}{2}$ the coins can have? _____

   Which coins would be in this half of the set? _____

6. What is the greatest value that $\frac{1}{2}$ the coins can have? _____

   Which coins would be in this half of the set? _____

7. Kent knows that $\frac{1}{4}$ of a set is less than $\frac{1}{2}$ of that set. So he's sure that $\frac{1}{4}$ of the coins *can't* be worth more than $\frac{1}{2}$ of the same coins. But Kelly disagrees. Who is right? Prove it! Explain with a picture and a caption.

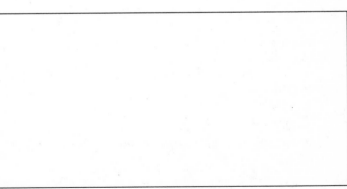

PREPARE TO SHARE ....

What did you find tricky about these problems? Explain on the back of this page.

FRACTIONS AND AGES

How can you use fractions to express ages?

GOAL: Students use fractions and/or mixed numbers to describe the ages of friends, family members, and pets.

MATERIALS: none

THE PLAN

1. Start this activity in class and have students complete it at home.

2. Begin by asking each member of the class to write down how old he or she is, in years and months (for example—10 years, 3 months). Guide students to round their age to the nearest year and month. Talk about ways to round. Then ask, "How old are you in years and fractions of a year?"

3. As needed, discuss how to name numbers of months as fractions of a year. You might give some examples to help students get started. For instance, write 9 years, 8 months on the chalkboard, and then elicit that this age translates to $9\frac{8}{12}$ years, or $9\frac{2}{3}$ years. Ask students how they would express 9 months, 3 months, 2 months, and so on, as fractions of a year. Then have students share their responses to your original question.

4. Assign students to gather the ages, in years and months, of friends, family members, neighbors, and pets, too! Have them record each age two ways in a table (like the one started below). Guide them to express all mixed numbers in simplest form.

Name of Person or Pet	Age in Years and Months	Age as a Mixed Number
Spot	3 years, 4 months	$3\frac{1}{3}$ years

5. Have students share their tables, as well as their experiences filling them in, in class the following day. Invite volunteers to post their charts.

FOLLOW-UP

⊗ Extend by having students use an almanac or other resource to find the ages of United States presidents at the time of their inauguration, the current ages of favorite TV actors, or the ages of famous composers, writers, artists, or athletes at the time of their greatest accomplishments or deaths, and so on. They should record these ages using mixed numbers in simplest form.

⊗ Challenge students to express the ages of the same family members and friends first in years and days and then in fractions of a year, in simplest form, based on that new data. Have them use 365 for the number of days in a year.

FRACTION ROLLERS, PART 2

How can you make the greatest mixed number given three digits? How can you make the least mixed number?

GOAL: Students play a game in which they create and compare mixed numbers.

MATERIALS: 1–6 number cubes (p. 85), index cards

THE PLAN

1. Have students form pairs. Provide each pair with a set of three number cubes. (Alternatively, they can roll one cube three times.) Tell students that they will play a game in which they will create mixed numbers by rolling the number cubes.

2. Present the rules for play:

   ✳ Each player rolls the number cubes to get three digits. Players should write the digits on scrap paper. The player uses those digits to form the greatest possible mixed number the digits can form. For example, with a roll of 2, 3, and 5, the greatest possible mixed number he or she can form is $6\frac{1}{2}$ ($5\frac{3}{2} = 6\frac{1}{2}$).

   ✳ The player with the greatest number for each round wins a point. No points are given for a tie.

   ✳ The first player to reach 10 points wins.

3. Vary the rules so that players try to make the least possible mixed number using the three digits rolled. In this case, the winner each time is the player with the lesser mixed number. Or vary the game to include three players; here, the winner can be the player with the greatest (or least) mixed number or the one whose number falls in the middle.

4. For whichever version of the game your students play, have them summarize the components of a winning strategy.

FOLLOW-UP

✳ Add some excitement by selecting a student "sportscaster" to offer a play-by-play of the action as pairs play. The sportscaster explains to the audience (other classmates) what players do and the strategies they use.

✳ Have students make up their own game in which they use cards, number cubes, spinners, or other manipulatives to compare mixed numbers. Guide them to think through their games and test them before writing the rules on index cards. Then have students present them to the class. Invite them to demonstrate their games. You may wish to have students contribute to a class book or folder of fraction games.

NOTING FRACTIONS

Did you know that standard musical notation involves the use of fractions?

GOAL: Students compute with fractions using musical notation as a code.

MATERIALS: student page 56

THE PLAN

1. Ask students whether they know how to read musical notation. Explain that in musical notation, the shape and color of a note indicates how long to sing or play it. For example, two half notes have the same duration as one whole note.

2. Copy the illustrated musical notation chart shown at right onto the chalkboard or chart paper. Invite knowledgeable volunteers to help you explain the standard notation of whole notes, half notes, quarter notes, eighth notes, and sixteenth notes. Help students identify the pattern in the progression from whole note to its fractional parts. Point out that the whole note looks like a clear oval. The half note has the same oval, but with a stem. The quarter note looks like the half note, but the oval is filled in. The eighth note looks like the quarter note, but has a small "flag" waving from the top of the stem. The sixteenth note looks like the eighth note, but has two "flags" on the stem.

Name	Note	Value
whole note	𝅝	1
half note	𝅗𝅥	$\frac{1}{2}$
quarter note	♩	$\frac{1}{4}$
eighth note	♪	$\frac{1}{8}$
sixteenth note	𝅘𝅥𝅯	$\frac{1}{16}$

3. Draw a series of musical notes on the chalkboard. Ask volunteers to envision the notes as a code that represents fractions. Guide them to write a numerical equation based on the musical phrase. Explain that when eighth notes or sixteenth notes are repeated, they may share "flags" like this:

4. Duplicate and distribute copies of page 56. Invite students to work on this page independently or in pairs.

FOLLOW-UP

⊛ Invite the school music teacher to follow up with further lessons on musical notation. Students might learn about the function of the dot (when a note is followed by a dot, the dot signifies augmenting the original note by $\frac{1}{2}$ its value; so, a dotted quarter note is worth $\frac{1}{4}$ + half of $\frac{1}{4}$, which is $\frac{1}{8}$).

⊛ Extend by having students create their own addition and subtraction equations using musical notation. Or have them translate standard fraction equations into musical notation.

NOTING FRACTIONS

Each musical note has a certain fractional value.

Use the chart to solve these equations. Rewrite the musical notes as a fraction sentence. Give each answer as a fraction or mixed number.

Name	Note	Value
whole note	𝐨	1
half note	♩	$\frac{1}{2}$
quarter note	♩	$\frac{1}{4}$
eighth note	♪	$\frac{1}{8}$
sixteenth note	♬	$\frac{1}{16}$

1) ♩ + ♩ $\quad \frac{1}{2} + \frac{1}{4} = \frac{3}{4}$ _____

2) ♩ + ♩ + ♫ _____

3) ♩ + ♪ + ♫ _____

4) 𝐨 + 𝐨 + ♩ + ♪ _____

5) ♩ + ♩ + ♩ + ♩ + ♪ _____

6) 𝐨 + ♪ + ♫ _____

7) ♩ + ♪ + 𝐨 + ♩ + ♪ _____

8) ♫ + ♫ + ♪ + ♫ _____

Now rewrite these standard fraction equations as musical phrases. Give the sum as a fraction or mixed number.

9) $\frac{1}{4} + \frac{1}{4} + \frac{1}{4} + \frac{1}{8} = $ ♩ + ♩ + ♩ + ♪

10) $\frac{1}{2} + \frac{1}{4} + \frac{1}{8} + \frac{1}{8} = $ _____

11) $1 + 1 + \frac{1}{2} + \frac{1}{16} + \frac{1}{16} = $ _____

12) $\frac{1}{2} + \frac{1}{2} + 1 + \frac{1}{2} + \frac{1}{4} = $ _____

13) $1 + 1 + \frac{1}{4} + \frac{1}{4} + \frac{1}{8} + \frac{1}{16} = $ _____

PREPARE TO SHARE

How would you write a note whose value is $\frac{1}{32}$? Explain on the back of this page.

A HEAD FOR FRACTIONS

Can you use mental math to find sums of fractions?

$$\frac{3}{5} + \frac{2}{5} = 1$$

GOAL: Students play a game in which they use mental math to find whole-number sums.

MATERIALS: 1–6 number cubes (p. 85) (or spinner, p. 90)

THE PLAN

1 Have students form groups of three. For each game, two students play and one is the judge. Provide each group with a pair of number cubes.

2 Introduce the rules of play:

⊗ Determine an order of play. Players will take turns rolling and adding.

⊗ The first player rolls the two number cubes, using the numbers shown to form a fraction—the lesser number is the numerator. (If a player rolls the same number on both cubes, he or she rerolls.)

⊗ The second player uses mental math to determine and then announce the fraction to add to that fraction to equal a sum of 1. If the player names the correct addend, he or she gets 1 point. No points are deducted for incorrect answers.

⊗ The judge can use either mental math or pencil and paper to check the answer. The judge also keeps score.

⊗ The first player to get 10 points wins.

3 After a game, students switch roles so that the judge becomes a player.

4 You can vary the game by changing the target sum. For example, have students try sums of 2 or 3, or of $1\frac{1}{2}$ or $2\frac{1}{2}$.

5 Ask students to explain the mental math strategies they used to find the sums. Encourage all sensible approaches. Invite students to demonstrate their strategies.

FOLLOW-UP

⊗ Challenge students by allowing them to use the number-cube rolls in any order to form the fractions. In this version of the game, students may have to subtract to reach the target sum.

⊗ Have students use three number cubes to form a mixed number in any way they wish. For example, a roll of 1, 3, and 6 can be interpreted as a mixed number such as $1\frac{3}{6}$, $3\frac{1}{6}$, $6\frac{1}{3}$, and so on. Then have students mentally add or subtract mixed numbers. For this version of the game, make a greater target sum. For instance, have players find sums of 5 or 6. Or challenge them to find sums like $7\frac{1}{2}$.

FRACTIONS IN ANCIENT EGYPT

Can you form fractions the way the ancient Egyptians did?

GOAL: Students use ancient symbols and rules to form fractions.

MATERIALS: student page 59

THE PLAN

1. Explain to students that several thousand years ago, the early Egyptians developed and used a system for expressing fractions. Copy the following ancient Egyptian fractions on the chalkboard or a poster.

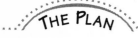

$$\frac{1}{4} \qquad \frac{1}{3} \qquad \frac{1}{5} \qquad \frac{1}{6} \qquad \frac{1}{10}$$

2. Have students examine these fractions and then practice writing them. Next, guide students to notice the pattern they follow, and help them describe it. Tell students that except for a few special fractions, all ancient Egyptian fractions used numerators of 1.

3. Write the following three exceptions on the board. Go over these exceptions with students. Again, have them practice writing the symbols for the fractions.

$$\frac{1}{2} \qquad \text{or} \qquad \frac{1}{2} \qquad\qquad \frac{2}{3} \qquad\qquad \frac{3}{4}$$

4. Tell students that when Egyptians formed fractions, they did not simply combine unit fractions. Show and discuss these two examples:

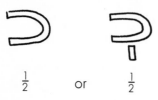  

5. Distribute copies of student page 59. Pair students to work together on the problems. Invite them to draw their solutions on the chalkboard. Allow time for students to explain how they followed the rules to form their fractions.

FOLLOW-UP

⊗ Invite pairs of students to form other fractions using Egyptian fraction rules.

FRACTIONS IN ANCIENT EGYPT

In ancient Egypt, fractions looked like this:

$\frac{1}{4}$ $\frac{1}{3}$ $\frac{1}{5}$ $\frac{1}{6}$ $\frac{1}{10}$

$\frac{1}{2}$ $\frac{2}{3}$ $\frac{3}{4}$

Example: $\frac{5}{12}$ =

Draw ancient Egyptian fractions for each of the following.

① $\frac{3}{4}$ =	② $\frac{1}{2}$ =
③ $\frac{2}{3}$ =	④ $\frac{5}{12}$ =
⑤ $\frac{7}{12}$ =	⑥ $\frac{5}{6}$ =

Mega-Fun Fractions ✪ Scholastic Professional Books

FLICKER FRACTION SUMS

Can your team flick its way to high sums?

Goal: Students strengthen their skills at adding fractions by playing a board game.

Materials: pennies or checkers, student pages 61–62 (rules and game board)

THE PLAN

1. Divide the class into teams of four students. Provide each team with a copy of the rules of play and a game board. Give each team a penny or checker to flick.

2. Explain that the object of the game is for each teammate to flick the penny (or checker) from its base on the playing board to achieve the greatest possible sum. Demonstrate how to do a proper flick. Urge students to flick the penny gently so that it doesn't fly in the air. Guide students to realize that the difficulty level for landing on the fractions varies.

3. Go over the rules together so that students understand them. Focus particularly on the rules that determine whether a coin has or has not landed on a fraction space.

4. Advise teams to keep a running total of their scores. The accuracy of their final sum can be checked by an impartial judge (you or a designated student or jury of students).

5. Have teams play several games during class. Teams can pair up and play against each other, in single elimination tournaments, or all teams can play at once and the highest-scoring team wins.

6. You may wish to keep the game in your math center for students to play as time allows.

FOLLOW-UP

⊛ Vary the game by having students create their own set of fractions for the board. Or invite them to change the size, shape, or number of the different regions on the board. You might even challenge them to change the shape of the board itself. For instance, they can make a 10-pointed star or an irregular shape.

⊛ Challenge students to vary the rules—for example, changing the number of flicks a player gets.

⊛ Challenge teams to get closest to without exceeding the target sum.

FLICKER FRACTION SUMS—RULES

PLAY: This game involves flicking a penny and adding fractions. You score when your penny lands totally inside one of the spaces. Your goal is to get the highest total score with all your flicks.

Here's how to play:

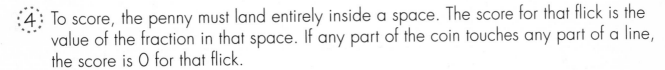

1. Put the game board on the floor or on a desktop. Place the penny inside the START space.

2. Players take turns flicking the penny. Always begin each new flick by putting the penny back in the START space.

3. Write down the fraction the penny lands on, then add that value to any previous score. Keep a running total for your team.

4. To score, the penny must land entirely inside a space. The score for that flick is the value of the fraction in that space. If any part of the coin touches any part of a line, the score is 0 for that flick.

5. Each player takes four turns in all. The team's final score is the sum of all valid flicks by all players.

6. When everyone on a team has had four flicks, the game ends for that team. Give your team's running score to the judge, who will check your addition.

That's it for the rules. Are you ready?
Choose an order of play for your team, and flick away.
Good luck!

PREPARE TO SHARE ···

How would you adjust the rules to make an even better game? Explain on the back of this page.

FLICKER FRACTION SUMS—GAME BOARD

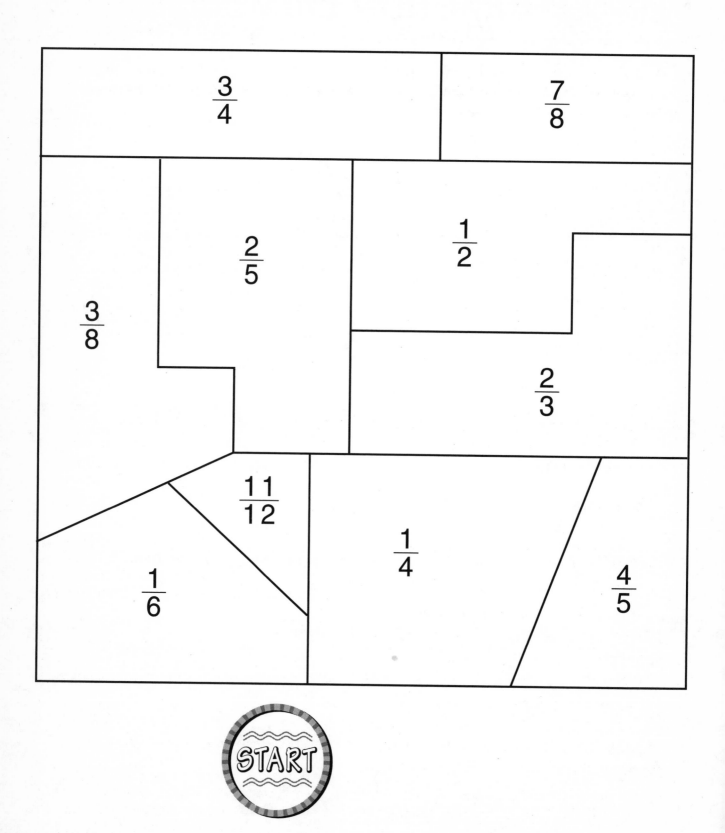

Mega-Fun Fractions ⊕ Scholastic Professional Books

FRACTION ADD-UP

Can you choose the addends if you know the sums of fractions and mixed numbers?

GOAL: Students strengthen fraction adding and subtracting skills by following rules to find addends that yield given sums.

MATERIALS: student page 64

THE PLAN

1. Tell students that in this activity, they must come up with addends to equal a sum they will already know. Remind them that they already know an appropriate strategy to use: work backward.

2. Ask students to imagine that they are in a health food store near a national park. Have them pretend that they are putting together a combination of foods to take along on a hike. Then present the following problem: "You want exactly 2 cups of trail mix made from peanuts, chocolate-covered raisins, and dried apple slices. You want different amounts of each ingredient, each amount less than 1 cup. How much of each ingredient do you get?"

3. Provide time for students to come up with their answers. Then ask volunteers to present their combinations and to explain what they did to arrive at those choices. Emphasize that many answers are possible, as long as the solutions follow the rules of the problem.

4. Duplicate and distribute copies of page 64, one per student. Go over the information on the page. Explain that the foods listed are commonly found in trail mixes, which are designed to provide hikers with maximum energy but minimum weight.

5. Emphasize the importance of reading each question carefully, as each situation places a different demand on the problem-solver. Remind students that questions have multiple solutions.

6. When students have finished, have them form small groups to compare answers and solution methods.

FOLLOW-UP

⊗ Have students formulate similar problems for partners to solve. Invite them to use all or some of the foods listed or to add some of their own to the collection.

⊗ Vary the activity by assigning weights, in fractions of a pound, to each of the foods. Challenge students to find combinations of foods that fit given weights and to make up questions using this new data.

⊗ As a challenge, assign prices—per pound, half-pound, and quarter-pound—to each item. Create basic trail mix problems (or have students create problems) that involve both quantities and total prices. Keep in mind that such problems require an intuitive understanding of multiplying fractions and money amounts.

FRACTION ADD-UP

Health Food Store

raisins	almonds
dried apple slices	yogurt-covered peanuts
dried cherries	
coconut flakes	pretzel nuggets
peanuts	sesame crisps
cashews	
chocolate-covered raisins	

Choose from the foods shown to make each trail mix blend described. Read each set of rules carefully. Give your answer as a labeled number sentence with fractions and/or mixed numbers. The first one has been done for you.

1. Make a 1-cup mix of almonds and pretzel nuggets. Use a different amount of each ingredient.

    $\frac{3}{4}$ cup almonds + $\frac{1}{4}$ cup pretzel nuggets = 1 cup mix

2. Make a 2-cup mix of three ingredients. Use $1\frac{1}{2}$ cups of yogurt-covered peanuts and two other foods. Use a different amount of each.

3. Make a 2-cup mix of four different ingredients. Use the same amount of two foods and different amounts of the remaining two. Use less than 1 cup of each food.

4. Make a 3-cup mix of four different ingredients. Use more than 1 cup of two foods and less than 1 cup of two others. Use different amounts of each ingredient.

5. Choose five ingredients. Use a different amount of each. Get enough to make more than 4 cups but less than 5 cups.

6. Choose a different amount of each of three ingredients. Then pick a different amount of each of three other ingredients so that both blends have the same number of cups.

Mega-Fun Fractions ⊗ Scholastic Professional Books

Sums on a Roll

How would you arrange four digits into two fractions to produce the greatest sum? The least sum?

GOAL: Students strengthen their number sense and skills at adding fractions and at comparing fractions and mixed numbers.

MATERIALS: 1–6 number cubes (p. 85)

THE PLAN

1. Have students form pairs. Provide each pair with two number cubes. Then tell them that they will play a game in which they create and add fractions.

2. Present the rules for play:
   - The goal of the game is to make a pair of fractions that produce the greatest sum.
   - Each player rolls the number cubes twice to get four digits. He or she records the digits and then uses them to form two fractions, each having a numerator that is less than its denominator. Here's an example: a student rolls 1, 5, 3, and 6. The greatest possible sum is $1\frac{1}{6} = \frac{5}{6} + \frac{1}{3}$.
   - The player with the greatest sum wins a point. No points are given for a tie.
   - The first player to reach 10 points wins.

3. You may wish to vary the rules in one or more of the following ways:
   - Have players try to make fractions that have the least sum or the greatest sum that does not exceed 1.
   - Have students play in groups of three. In this case, the winner can be the player with the greatest sum, the least sum, or the one with the sum in the middle.
   - Allow players to form fractions in any way they wish, not only with smaller numerators.

4. Whatever version of the game students play, remind them how to express improper fractions as mixed numbers.

5. Emphasize that finding the greatest (or least) sum is likely to be a trial-and-error process. Encourage students to communicate winning strategies as they play.

FOLLOW-UP

- As an alternative, post a set of fractions with different denominators on an interactive bulletin board. Students can rearrange the terms of two of these fractions to make a pair of new fractions that will have a greater (or lesser) sum than the original.

- Extend the game by having students play with a 1–8 spinner or 1–10 number cards.

FRACTION PATH PUZZLES

Can you complete a puzzle that involves adding and subtracting fractions?

GOAL: Students use logical reasoning, number sense, and their understanding of adding and subtracting fractions to complete three fraction path puzzles.

MATERIALS: student page 67

THE PLAN

1. Write the following incomplete number sentence on the board: $\frac{1}{2} \ \square \ \frac{1}{4} = \frac{3}{4}$. Ask students to determine the operation sign that belongs in the box (+) and to give their reasons.

2. Tell students that you are going to give them three different fraction path puzzles that are missing the operation signs for addition (+) and subtraction (−).

3. Have students work individually or in pairs. Duplicate and distribute page 67. Go over the path puzzles together to ensure that students understand how to approach them. Point out that each puzzle is missing operation signs; students must decide which sign goes where so that the string of computations from START to the shaded box works to yield the final number, 0.

4. You may want to provide students with some hints to help them complete this task, which they may find challenging. Possible suggestions include:

   ⊗ Guess and check; revise as necessary.
   ⊗ Work backward from the final number.
   ⊗ Rewrite all fractions in equivalent form, all with the same denominator.
   ⊗ Use manipulatives.

FOLLOW-UP

   ⊗ As a simpler alternative, present similar but shorter fraction path puzzles.
   ⊗ Extend by having students create their own fraction path puzzles.

FRACTION PATH PUZZLES

Can you complete the fraction puzzle path?

Write a + or − sign in each blank square to form a fraction path that ends at 0.
Check your work on scrap paper.

START

| $\frac{1}{4}$ | | $\frac{1}{2}$ | | $\frac{3}{4}$ | = | 0 |

START

| $\frac{1}{2}$ | | $\frac{2}{3}$ | | $\frac{5}{6}$ |

| 2 | = | 0 |

START

| $\frac{1}{2}$ | | $\frac{2}{3}$ |

| $\frac{1}{4}$ | | $\frac{5}{6}$ |

| $\frac{7}{12}$ | = | 0 |

⋯⋯⋯ **PREPARE TO SHARE** ⋯⋯⋯⋯⋯⋯⋯⋯⋯⋯⋯⋯⋯⋯⋯⋯⋯

How did you decide which operation to use?

FRACTION MAGIC FIGURES

Can you complete puzzles that involve adding and subtracting fractions?

Goal: Students use logical reasoning, number sense, and their understanding of adding and subtracting fractions to complete four magic figures puzzles.

Materials: student page 69

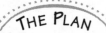

THE PLAN

1. Write the following incomplete number sentence on the board: $\frac{1}{2} + \boxed{} = \frac{3}{4}$. Ask students to determine the missing fraction that would correctly complete the number sentence, and to give their reasons. $(\frac{1}{4})$

2. Review the concept of the magic square. Most students will have had experience with these common puzzles, in which the sum of all numbers in every row, column, and diagonal is the same—the magic number.

3. Tell students that you will give them four different fraction magic figures puzzles—one triangle and three different squares. Guide them to notice that each puzzle is missing some of the fractions (and/or mixed numbers) that will complete it.

4. Have students work individually or in pairs. Duplicate and distribute page 69. Go over the puzzles together to ensure that students understand how to approach them. Point out that students must determine the missing numbers that will make the sum of each row, column, and diagonal yield the magic number. Be sure students understand that each figure has its own unique magic number.

5. You may want to provide students with some hints to help them complete this task, which they may find challenging. Possible suggestions include:

   ⊗ Guess and check; revise as necessary.

   ⊗ Work backward from the magic number.

   ⊗ Rewrite all fractions in equivalent form, all with the same denominator.

   ⊗ Use manipulatives.

FOLLOW-UP

⊗ As an alternative, present simpler (or harder) fraction magic squares puzzles.

⊗ Extend by having students create their own fraction magic squares puzzles.

⊗ Challenge students to solve a fraction magic square puzzle in which no row, column, or diagonal is complete; students must determine the magic sum on their own.

Name _____ Date _____

FRACTION MAGIC FIGURES

Follow the directions to fill in the four fraction figures. But be careful!
These figures can be tricky!

1 Write the numbers $1\frac{1}{3}$, $1\frac{2}{3}$, 2, $2\frac{1}{3}$, $2\frac{2}{3}$, and 3 in the circles
so that the sum of the three numbers along each side of the
figure equals the number written inside the figure.

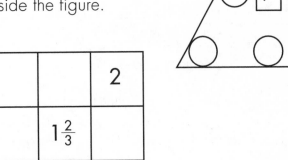

2 Use the numbers $\frac{1}{3}$, 1, $2\frac{1}{3}$,
$2\frac{2}{3}$, and 3 to complete the
magic square. The sum of the
three numbers in each row, in
each column, and in each
diagonal must be 5.

		2
	$1\frac{2}{3}$	
$1\frac{1}{3}$		$\frac{2}{3}$

3 Complete this magic square.
Use the numbers $1\frac{1}{2}$, 3, $4\frac{1}{2}$,
and 12. Make the magic sum
the same in each row, column,
and diagonal.

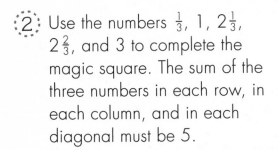

		6
	$7\frac{1}{2}$	$13\frac{1}{2}$
9	$10\frac{1}{2}$	

4 Complete this magic square.
Use the numbers $\frac{5}{8}$, 1, $1\frac{1}{4}$, $1\frac{3}{8}$, $1\frac{1}{2}$, $1\frac{3}{4}$,
and $1\frac{7}{8}$. Make the magic sum the same
in each row, column, and diagonal.
(HINT: First figure out the magic sum.)

2	$\frac{1}{4}$	$\frac{3}{8}$	$1\frac{15}{24}$
$\frac{9}{8}$	$\frac{7}{8}$	$\frac{3}{4}$	
$\frac{1}{2}$			$\frac{1}{8}$

PREPARE TO SHARE

How did you begin to solve these problems? Explain your strategy on the back of
this page.

PARTS OF PARTS

How can you model what it means to multiply fractions?

GOAL: Students use grid paper and counters to visualize the concept of multiplying two fractions, leading to an understanding of the algorithm.

MATERIALS: centimeter grid paper (p. 88), scissors, two colors of counters, student page 71

THE PLAN

1. Duplicate and distribute sheets of grid paper to students. Have them use scissors to cut out a 4 x 4 and a 6 x 6 grid from the paper.

2. Have them place the 4 x 4 grid on the center of their work area and use counters of one color to cover $\frac{1}{2}$ the grid. Ask: "How many squares did you cover?" (8) Now have students use the second color of counters to cover $\frac{1}{2}$ of that half. Ask: "How many squares have two counters on them? (4) What fraction of the whole grid is covered with two counters?" ($\frac{1}{4}$)

3. Ask a volunteer to write a number sentence that fits what you just modeled: $\frac{1}{2} \times \frac{1}{2} = \frac{1}{4}$. Reiterate what students modeled—they found a fraction of a fraction. Restate the meaning of the related number sentence in words: Half of one-half is one-fourth.

4. Now have students use the 6 x 6 grid and the counters to model $\frac{1}{6} \times \frac{1}{2}$. Have them first cover $\frac{1}{6}$ of the grid with one color of counters, then cover $\frac{1}{2}$ of that $\frac{1}{6}$. Record the model in a number sentence: $\frac{1}{6} \times \frac{1}{2} = ?$ ($\frac{1}{12}$)

5. Duplicate and distribute page 71 for students to work on individually or in pairs.

FOLLOW-UP

⊗ Summarize the activity by having students write complete number sentences to match their models. Encourage students to look for a pattern in the number sentences that can lead them to derive an algorithm for multiplying fractions. (*Multiply the numerators, multiply the denominators, simplify the product, if possible.*)

⊗ Extend the activity by having students model similar fractions of fractions on 8 x 8 or 10 x 10 grids. With larger grids they can model a greater variety of fractional parts.

⊗70

PARTS OF PARTS

Use a 4 x 4 grid and two colors of counters to model these fractions.

1 Cover $\frac{1}{2}$ of the grid with one color counter.
Cover $\frac{1}{4}$ of that $\frac{1}{2}$ with the other color counter.
What fraction of the whole grid is covered with two counters? _____

2 Cover $\frac{1}{4}$ of the grid with one color counter.
Cover $\frac{1}{2}$ of that $\frac{1}{4}$ with the other color counter.
What fraction of the whole grid is covered with two counters? _____

3 Cover $\frac{3}{4}$ of the grid with one color counter.
Cover $\frac{1}{2}$ of that $\frac{3}{4}$ with the other color counter.
What fraction of the whole grid is covered with two counters? _____

Use a 6 x 6 grid and two colors of counters to model these fractions.

4 Cover $\frac{1}{2}$ of the grid with one color counter.
Cover $\frac{1}{3}$ of that $\frac{1}{2}$ with the other color counter.
What fraction of the whole grid is covered with two counters? _____

5 Cover $\frac{1}{3}$ of the grid with one color counter.
Cover $\frac{1}{2}$ of that $\frac{1}{3}$ with the other color counter.
What fraction of the whole grid is covered with two counters? _____

6 Cover $\frac{2}{3}$ of the grid with one color counter.
Cover $\frac{1}{2}$ of that $\frac{2}{3}$ with the other color counter.
What fraction of the whole grid is covered with two counters? _____

PREPARE TO SHARE ....

Can you develop a rule for multiplying fractions? Explain on the back of this page.

ROLL, ROUND, AND RECORD

Can you correctly multiply a fraction or mixed number by a whole number and then round the product to the nearest whole?

GOAL: Students play a game to practice multiplying fractions and mixed numbers and rounding products.

MATERIALS: student page 73, 1–6 number cube (p. 85)

THE PLAN

(1) Have students form small groups to play this multiplying and rounding game. Duplicate and distribute page 73. Notice that the page has two games—one with fractions in all "lanes," the other with fractions and mixed numbers. Give each group a number cube.

(2) To begin, players pick a lane to use for the game and write their name in the appropriate space.

(3) Here's how to play:

⊗ In turn, each player rolls the number cube and multiplies that number by the first fraction (or mixed number) in his or her lane. The player then rounds the product to the nearest whole number and records the rounded number on a separate score sheet.

⊗ Each player checks the math of the preceding player.

⊗ In turn, other players roll, multiply, round, and record their score for the first fraction (or mixed number) in their lane on the score sheet. In subsequent turns, players add the new rounded number to the preceding score to keep a running total as they multiply their way across their lanes.

⊗ The player with the highest total after completing all turns in the "race" wins.

FOLLOW-UP

⊗ Vary the game by using a number cube or spinner with numbers greater than 1–6.

⊗ Vary the game by changing the numbers in the lanes. Include decimals.

Roll, Round, and Record

Choose a lane. You will need a number cube.
Roll, multiply, round, and record as you race across your lane.
Keep a running score. Who will win?

Game 1

Player	1	2	3	4	5
	$\frac{1}{2}$	$\frac{3}{8}$	$\frac{3}{5}$	$\frac{1}{4}$	$\frac{5}{8}$
	$\frac{2}{3}$	$\frac{1}{5}$	$\frac{1}{2}$	$\frac{4}{5}$	$\frac{3}{8}$
	$\frac{1}{2}$	$\frac{5}{6}$	$\frac{3}{4}$	$\frac{1}{5}$	$\frac{4}{5}$
	$\frac{3}{4}$	$\frac{2}{5}$	$\frac{1}{8}$	$\frac{5}{6}$	$\frac{1}{2}$

Game 2

Player	1	2	3	4	5	6
	$\frac{1}{5}$	$2\frac{1}{2}$	$\frac{3}{8}$	$1\frac{1}{4}$	$\frac{5}{6}$	$1\frac{1}{2}$
	$\frac{5}{6}$	$\frac{1}{8}$	$1\frac{4}{5}$	$1\frac{2}{3}$	$\frac{3}{5}$	$3\frac{1}{4}$
	$\frac{2}{3}$	$1\frac{1}{2}$	$\frac{7}{8}$	$2\frac{1}{4}$	$1\frac{5}{6}$	$\frac{1}{2}$
	$\frac{3}{5}$	$\frac{1}{8}$	$1\frac{3}{4}$	$1\frac{1}{3}$	$\frac{3}{4}$	$2\frac{2}{3}$

Prepare to Share

Does it matter which lane you choose? Explain your thinking on the back of this page.

FRACTIONS AND CALORIES

How can fractions help you figure out the number of calories you burn?

GOAL: Students use their understanding of fractions of an hour to interpret information in two charts and to formulate problems based on the given data.

MATERIALS: student page 75

THE PLAN

1. Explain the meaning of *calorie* (a unit of heat energy supplied by food). Brainstorm with students some foods that are high in calories (pizza, cheeseburgers, and ice cream) and low in calories (vegetables, fish, and fruit). Then discuss the principle that the body burns, or uses up, calories as it performs activity. Point out that the number of calories burned varies with the activity (an active game of handball burns more calories than a seated game of checkers) and is also affected by how long we do the activity (a 1-hour walk burns more calories than a 10-minute walk).

2. Duplicate and distribute page 75 to pairs of students. Go over the data presented. To help students understand the data, ask questions, such as "Which exercise burns the greatest number of calories per hour? Which exercise burns about three times as many calories as walking does?"

3. Focus students' attention on the Calories in Some Snack Foods chart. Help them read and analyze the information presented by asking questions: "Which food listed is highest in calories? Which is lowest in calories? Does any of the data surprise you?"

4. Direct students to answer the questions. Help them give each answer in minutes and as a fraction (or mixed number) of an hour (in simplest form).

5. Discuss answers together when students have finished. Talk about any discrepancies.

FOLLOW-UP

⊗ Extend by helping students expand the top table to include other exercises or everyday activities (strolling, typing on the computer, raking leaves, and so on) and the rate of calories burned for each. Suggest sources where students can find this data.

⊗ Invite students to add foods—and their calorie counts—to the second table.

⊗ Challenge students to find the time it would take to burn greater or lesser amounts of each food. Guide them to use mental math (and their intuitive understanding of proportional reasoning) to figure out these answers. Ask them to explain their reasoning.

FRACTIONS AND CALORIES

How many calories do you burn when you exercise?

A 100-pound person burns the following number of calories per hour of each activity:

	Biking	Running	Swimming	Volleyball	Walking
Activity					
Calories Per Hour	320	500	350	125	150

Calories in Some Snack Foods

hot dog	150 calories
plain cookie	55 calories
8 tortilla chips	80 calories
2 Tbsp. peanut butter	200 calories
1 can regular soda	150 calories
$\frac{1}{2}$ cup ice cream	200 calories
slice of pound cake	250 calories

Use the data from the two charts to answer these questions about Ann, who weighs 100 pounds.

1. Ann eats a hot dog. How long must she walk to burn those calories? _____

2. Ann eats 8 tortilla chips. How long must she bike to burn those calories? _____

3. Ann eats 2 tablespoons of peanut butter. How long must she run to burn those calories? _____

4. Ann eats 3 cookies. Which activity must she do for about a half-hour to burn off the calories? Explain. _____

5. Ann has a slice of cake and a can of soda. What fraction of the calories will she burn if she swims for an hour? Explain. _____

_____

6. Ann takes a 4-hour walk. How much ice cream can she eat and not have any calories left over? Explain. _____

_____

FRACTION STORIES

Can you create story problems that you can solve by using fractions?

GOAL: Students write and illustrate story problems involving operations with fractions.

MATERIALS: binder, paper, crayons

THE PLAN

1. Read the following aloud to students: "A train is 9 cars long. Two-thirds of the cars carry passengers. One-third of the cars carry freight. How many cars are passenger cars?" (6)

2. Ask students to explain how they would use fractions to solve this problem. Also ask them to make a sketch to illustrate the problem and its solution.

3. Guide students to recognize that they multiplied by a fraction $(9 \times \frac{2}{3})$ to solve the train problem. Then present problems they can solve by adding or subtracting fractions. For example: "Gene ate $\frac{1}{2}$ of the cake after dinner. Before he went to sleep, he ate another $\frac{1}{3}$ of it. What fraction of the cake did Gene eat in all?" Or: "An ant has climbed $2\frac{1}{4}$ feet up the leg of a picnic table piled high with treats. If the tabletop is $3\frac{3}{4}$ feet high, how much farther must the ant climb to get at the goodies?" Discuss students' solutions to these problems, and invite them to illustrate the problems and solutions.

4. Tell students that they will create fraction story problems. Ask each student to write three (or more) fraction stories that you will collect in a binder entitled "Fun Fraction Stories." Have students write one problem per page, accompanied by a sketch of it. Ask them to provide the answer on a self-stick note.

5. Encourage students to make their fraction story problems about real-life activities, using actual or made-up data. Stories can be short or long and may consist of multistep problems to be solved by understanding, adding, subtracting, or multiplying fractions.

6. Collect students' completed stories and bind them. Create a solution page in the back. Display the book for students to use. Invite additional contributions at any time.

FOLLOW-UP

✴ Challenge students to write problems that involve more than one concept, such as fractions and money, fractions and measurement, or fractions and geometry.

✴ Display some of the best story problems or those particularly well illustrated.

✴ Have students create an online fraction storybook.

FRACTION SCAVENGER HUNT

How clever are you at finding fractions around you?

GOAL: Students make visual estimates and then measure to locate items that fit given descriptions.

MATERIALS: student page 78, rulers

THE PLAN

1. Explain what a scavenger hunt is—a game that involves searching for items on a given list. Have students work in pairs. Decide whether to restrict the search to the classroom or to extend it to include other parts of the school environment.

2. Duplicate and distribute page 78 to each pair of scavengers. Provide rulers. Take a few minutes to go over the items on the search list. Then set a reasonable time limit for pairs to work their way down the list to find an example of each item on it. Point out that students need not search for items in the order given. Explain that when the time runs out, the winning pair will be the one that has found the greatest number of items.

3. As needed, talk about how to make estimates. You might introduce the idea of using a benchmark as an estimating aid. For example, knowing that a sheet of notebook paper is 11 inches long may help students estimate lengths of 1 foot. Encourage pairs to come up with their own benchmarks to guide them in their hunt.

FOLLOW-UP

⊗ When the time is up, have pairs post their lists, share their findings, and compare notes about the hunt. Invite them to tell what about the search was easiest and most difficult. Encourage students to share insights they gathered during the activity.

⊗ Extend by having students search for items that weigh a particular amount, perhaps $\frac{1}{4}$ pound or $2\frac{1}{2}$ pounds. Or challenge them to brainstorm a list of things that take a certain length of time to do—for example, $\frac{1}{2}$ minute, $\frac{1}{4}$ hour, or $1\frac{1}{2}$ hours.

FRACTION SCAVENGER HUNT

You will need a ruler and a pencil. Work in pairs to fill in the table as quickly as you can. Measure accurately! How many items can you find?

Description of Item	What We Found
1. Something about $\frac{1}{2}$ inch long	
2. Something about 2 inches long	
3. A pencil about $\frac{1}{3}$ of a foot long	
4. Something about $1\frac{1}{2}$ feet long	
5. A tool about $\frac{3}{4}$ of a foot long	
6. Something about $\frac{2}{3}$ of a yard in length	
7. A drawer about a third of a foot high	
8. Something between 2 and $2\frac{1}{2}$ feet high	
9. A box about $2\frac{1}{2}$ inches wide	
10. A book about $3\frac{1}{2}$ inches thick	
11. Something about half your height	
12. Something about 5 feet from a door	
13. A cabinet or shelf about $3\frac{1}{2}$ feet tall	
14. Something $1\frac{1}{2}$ times as long as it is wide	
15. A shoe less than 1 foot long	
16. Two people, one of whom is about $4\frac{1}{2}$ inches taller than the other	

PREPARE TO SHARE

What did you like best about this activity? What was hardest? Explain on the back of this page.

ONLY ONE-THIRD AGREED THAT...

How can you use fractions to describe the results of a survey?

GOAL: Students poll classmates on a topic of their choice and then interpret the results.

MATERIALS: none

THE PLAN

1. Discuss with students what a survey is and how and by whom surveys are used. Talk about the usefulness of surveys and, if appropriate for your class, the problems inherent in them. If possible, show students an example of a survey, perhaps one you received in the mail or one in a local newspaper. Invite students who have participated in surveys to describe their experiences.

2. Then have students form small groups. Have each group brainstorm a list of topics on which to survey classmates. Guide them to consider topics of interest to them, from favorite TV programs, music groups, or sports, to their views on environmental, civic, or legal issues. One student in each group will act as secretary and record its survey questions.

3. Point out that questions within a poll need not be related. Once groups have settled on a list of topics to pursue, guide them to write the specific questions they would like to ask. Direct them to word each question so that it is easy to understand. Guide them to provide a selection of short answers that can be easily recorded. For example, questions about favorite things could list 3–5 reasonable choices from which all respondents must select. Questions about issues of interest to students can call for answers like Yes or No, True or False, Agree or Disagree, I Support or I Oppose.

4. Provide time for groups to prepare and edit their survey questions. Then collect the questions, duplicate them, and distribute them to each student in the class.

5. Allow time for students to answer the survey questions thoughtfully. Then collect all the forms and give them back to the group that generated each so that members can see how their classmates responded to their questions.

6. Talk with students about ways to use fractions to describe the results of their surveys. Then give groups time to do a fractional analysis.

7. Have groups summarize and present their findings to the class. Guide them to use fractions in their analyses as much as possible.

FOLLOW-UP

⊗ Talk about what the results of all the surveys tell students about their views and about themselves. Talk about what students have learned from the process, discussing ways that using fractions is helpful. Then talk about ways to improve on all aspects of the polling process for the next time.

FRACTIONS EVERY DAY

How do people use fractions in their daily lives?

GOAL: **Students interview people to learn what they know about fractions and how they use them.**

MATERIALS: **fraction surveys (see below), clipboards (optional), tape recorders (optional)**

 THE PLAN

1. Ask students to guess what people they know will say when asked questions about the meaning of fractions or about how they use fractions in their everyday lives. Then tell students that they are going to undertake an activity to find out these answers.

2. As a class, generate a list of questions to ask family members, friends, and members of the school community. Here are some examples; many others are possible:

   ⊗ What is a fraction?

   ⊗ Do you ever add, subtract, multiply, or divide with fractions? Explain.

   ⊗ How is a fraction like a decimal?

   ⊗ What is the hardest thing to understand about fractions? What is the easiest?

   ⊗ How do you use fractions at work? At play?

   ⊗ How do you use fractions when you shop? When you cook? When you sew?

   ⊗ How do you use fractions when you do household chores?

   ⊗ Where do you see fractions outside the house?

3. Combine the best of students' questions into a fraction questionnaire. Make several copies of the questionnaire for each student. Then assign student pollsters to interview various people in their neighborhood—people of all ages and backgrounds, if possible—to learn how people and fractions get along. Provide clipboards and tape recorders, if available.

4. Back in class, help students analyze and compare the answers to the questionnaires. After students share their findings, summarize together what the survey reveals about people and fractions. Students may wish to share their findings with those people they interviewed.

 FOLLOW-UP

⊗ Extend by having students prepare a letter to the editor of the local newspaper in which they describe the questionnaire they created and the results it showed. In the letter, they can give their opinion on the value of having and using knowledge about fractions.

HALVES ALL AROUND

Students identify all the everyday ways they use the concept of one-half.

MATERIALS: none

⊗ Elicit from students what is meant by the concept of one-half. List their responses on the chalkboard. Ask them to predict what time it will be when half the school day is over.

⊗ Then have students form small groups. Have groups brainstorm a list of all the different ways they use one-half in their lives. Encourage them to think about things they customarily divide in half (oranges, sandwiches), places they see the concept used (sporting event halftimes, advertisements for sales), things they do half at a time (eat a sandwich).

⊗ Ask groups to share their ideas in a whole-class discussion. Invite them to give demonstrations as needed. Tell them when you are about halfway through the activity.

LOOKS LIKE A FRACTION TO ME

Students estimate and then find fractions based on volume.

MATERIALS: cylindrical glass, water, centimeter ruler

⊗ One student fills a container with water. Another estimates the fraction of the glass that is filled. A third measures the actual fraction.

⊗ Students take turns filling and estimating.

⊗ Demonstrate how to use the ruler to measure the fraction.

FRACTION GARDENS

Students use an understanding of fractions of a region to plan a garden.

> MATERIALS: **centimeter grid paper (p. 88), crayons**

⊗ Students outline a large rectangular region of the grid paper to use as their garden.

⊗ They mark off patches of that region and color and label each with the name of a kind of tree, plant, flower, or vegetable.

⊗ Then they write a fraction to show what part of the entire garden each patch is.

⊗ Repeat with gardens and patches of different sizes. Include irregular polygons and other shapes.

FRACTION FORMATIONS

Students manipulate digits to find different fraction sums and differences and to form equivalent fractions.

> MATERIALS: **four number cards (one each for the numbers 1, 2, 4, and 8)**

⊗ Have pairs of students use the cards—one per each term of a fraction—to create two fractions with the greatest (or least) possible sum.

⊗ Have pairs form two fractions with the greatest (and least) difference.

⊗ Have pairs use the four cards to make the greatest (and least) single fraction with a two-digit numerator and a two-digit denominator. Have them use any of the cards to form a fraction closest to or equivalent to a fraction you name, such as $\frac{1}{4}$ or $\frac{2}{3}$.

FRACTION FLAGS

Students examine and analyze world flags to look for fractional patterns.

MATERIALS: pictures of various world flags
(from almanacs, encyclopedias, or Web sites)

✲ Provide pictures of world flags students can examine and analyze to determine which ones are divided into halves, thirds, fourths, or other fractional regions.

✲ Help students develop a Fraction Flags bulletin board that displays flags sorted by fractional patterns. For example, they might include the flags of Italy and Gabon to show thirds, and use the flags of Monaco and Indonesia to suggest halves.

✲ Have students include flags that do not readily represent clear fractional parts, such as Kiribati, Qatar, Seychelles, and Zimbabwe.

✲ Challenge students to use equivalent fractions to describe the fractional parts of the flags of Eritrea, Chile, Jamaica, and Thailand.

FRACTION RECIPES

Students apply their knowledge of fractions to work with recipes.

MATERIALS: recipes that involve fractional amounts (see cookbooks or Web sites)

✲ Provide recipes. Challenge students to determine the quantities to use to prepare double or triple the amount, or to divide the recipe in half or in fourths.

✲ Have a cooking day on which small groups prepare simple recipes together.

FRACTION STRIPS

Number Cubes

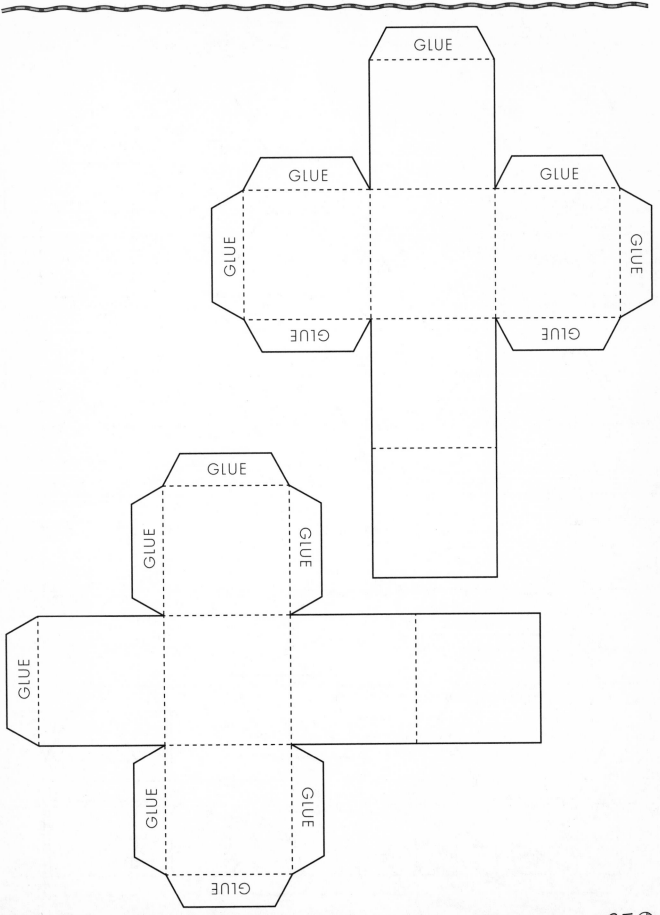

Pattern Blocks

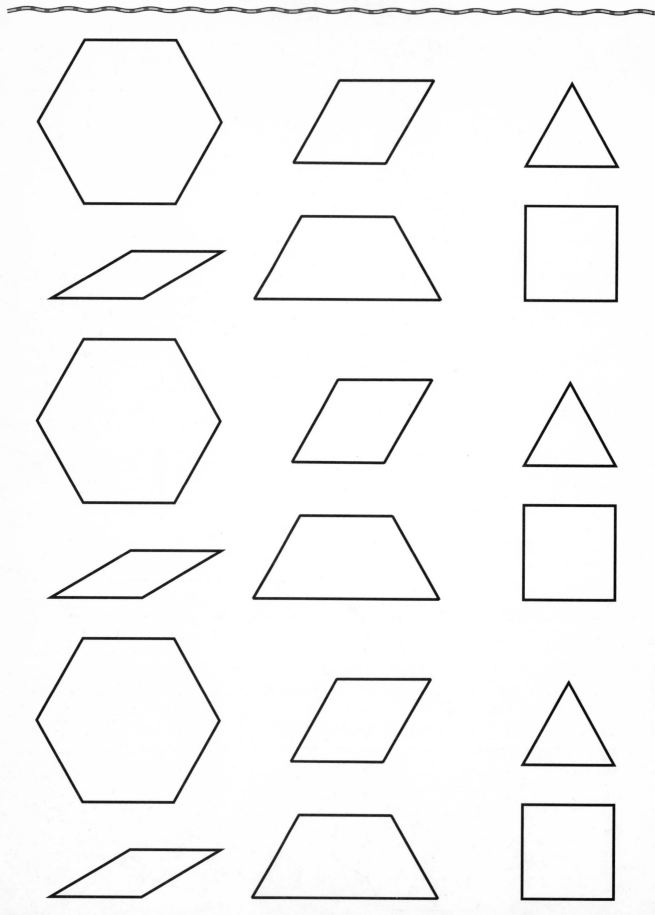

Dot Paper

Centimeter Grid Paper

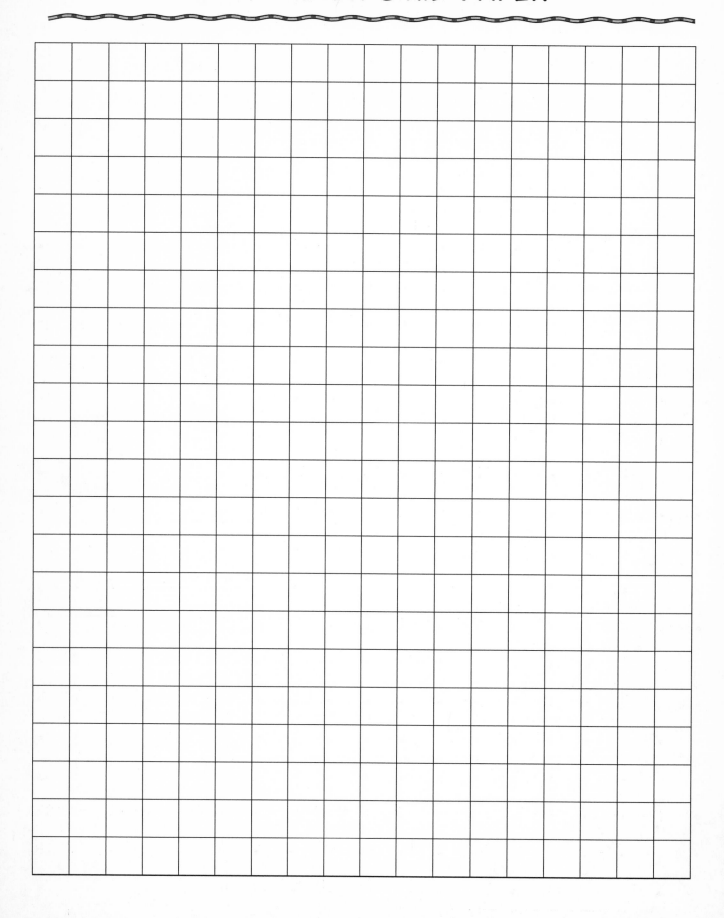

INCH GRID PAPER

SPINNER

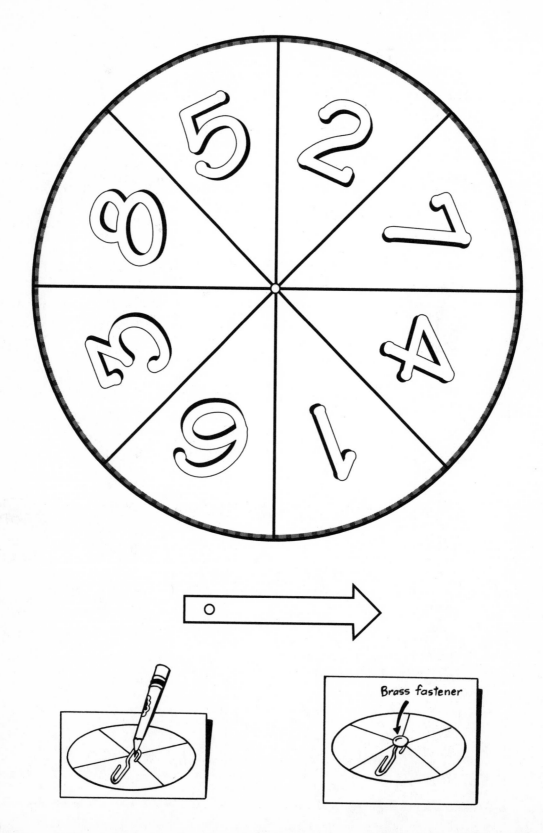

Brass fastener

Mega-Fun Fractions ⊗ Scholastic Professional Books

COINS

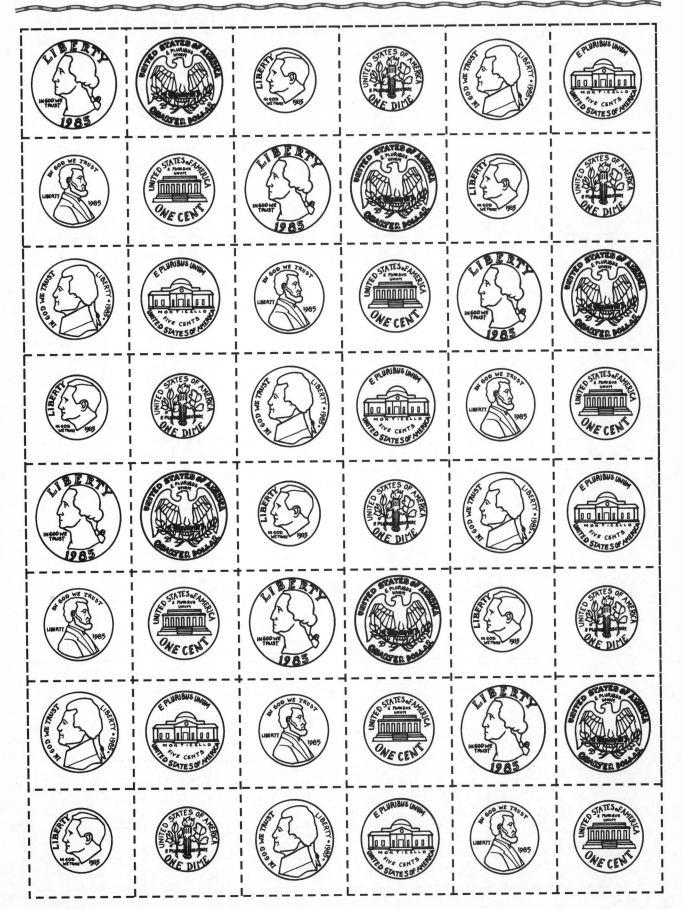

Name _____ Date _____

FRACTION SELF-EVALUATION FORM

(1) Here's how I feel about fractions:_____

_____

_____.

(2) My favorite fraction activity was _____

_____.

I liked it because _____

_____.

(3) The fraction activity that was hardest for me was _____

_____.

I think it was hard because _____

_____

_____.

(4) An important idea I learned about fractions is_____

_____

_____.

(5) I rate my understanding of fractions as (circle one)

      Strong   Good   Fair   Weak

Explain. _____

_____.

(6) Something I still don't understand about fractions is _____

_____

_____.

(7) In future work with fractions, I hope to learn_____

_____

_____.

Mega-Fun Fractions ⊗ Scholastic Professional Books

HALFNESS
(p. 8)

Answers will vary.

FILL ALL FOUR
(p. 11)

Solutions will vary.

WHAT'S IN PETAL, MISSISSIPPI?
(p. 12)

The Checker Hall of Fame

WHAT'S LEFT?
(p. 13)

1. $\frac{1}{4}$
2. a. $\frac{1}{2}$; b. $\frac{3}{4}$; c. $\frac{1}{4}$
3. a. $\frac{1}{3}$; b. $\frac{5}{6}$; c. $\frac{1}{6}$
4. a. $\frac{7}{8}$; b. $\frac{1}{8}$; c. $\frac{5}{8}$

FRACTIONS UP A TREE
(p. 14)

Check students' drawings.

COLORFUL REGIONS
(p. 15)

Check students' drawings.

FRACTION CHART PUZZLE
(p. 17)

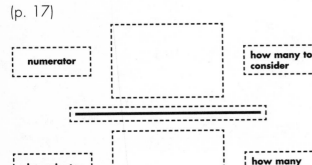

numerator

how many to consider

denominator

how many in all

PART ART
(p. 18)

Check students' figures.

COLLABORATIVE QUILTS
(p. 19)

Quilts will vary.

PATTERN BLOCK PROOFS
(p. 21–22)

1. $\frac{1}{8}$; $\frac{2}{8}$ or $\frac{1}{4}$
2. $\frac{3}{8}$; $\frac{3}{8}$
3. $\frac{1}{8}$; $\frac{3}{8}$
4. $\frac{1}{32}$; $\frac{6}{32}$ or $\frac{3}{16}$
5. $\frac{6}{32}$ or $\frac{3}{16}$; $\frac{18}{32}$ or $\frac{9}{16}$
6. Answers will vary.

SHADED SHAPES
(p. 24)

Answers will vary. Check students' drawings.

STAND UP FOR FRACTIONS
(p. 25)

Answers will vary with the number of students in your class.

CLASS FRACTION WALL
(p. 26)

Answers will vary with the number of students in your class.

PICTURE THESE FRACTIONS
(p. 27)

Answers will vary according to the photographs displayed.

MIXING SNACK MIX
(p. 29)

1. 12 raisins, 6 nuts
2. 18 nuts, 6 chocolate chips
3. 12 raisins, 12 chocolate chips, 12 nuts
4. No; there are 3 too few chocolate chips
5. Sample answer: $\frac{1}{2}$ are chocolate chips, $\frac{3}{8}$ are raisins, $\frac{1}{8}$ are nuts
6. Recipes will vary.

FRACTION MESSAGE
(p. 30)

Albuquerque, New Mexico

THE LANGUAGE OF FRACTIONS
(p. 31)

1. numerator
2. fifth
3. sixth
4. quarter
5. improper
6. half
7. equivalent
8. denominator
9. eighth
10. tenth
11. third
12. fourth

FRACTION DICTATION
(p. 32)

$\frac{1}{2}, \frac{2}{7}, \frac{1}{4}, 2\frac{1}{2}, \frac{2}{3}, 2\frac{1}{5}, \frac{5}{6}, \frac{1}{3}, \frac{7}{8}, \frac{11}{12}, 4\frac{1}{4}, \frac{98}{100}$

FRACTIONS AND EGG CARTONS
(p. 33)

Check students' models.

BADGE BUDDIES
(p. 34

Check the badges on each buddy pair.

EQUIVALENT FRACTION CONCENTRATION
(p. 35)

Check students' pairs as they play.

FRACTION BINGO—TIMES TWO
(p. 36)

Winning cards will vary.

FRACTION POEMS
(p. 40)

(No answer)

SHARING FRACTION PIE
(p. 41)

Check students' models.

FRACTION FILL 'EM UP
(p. 42)

Check students' models.

PROVE IT!
(p. 43)

Evaluate students' arguments as they are presented.

FRACTION WAR
(p. 44)

Observe students as they play.

FRACTION ROLLERS, PART 1
(p. 45)

Greatest and least fractions will vary depending on the numbers rolled.

THAT'S AN ORDER!
(p. 46)

Answers will vary depending on the cards drawn. Observe students as they play.

FRACTIONS OF A DAY
(p. 47)

Answers will vary for each student.

TIME FOR FRACTIONS
(p. 50)

Observe students as they play.

COINING FRACTIONS
(p. 51)

Answers will vary depending on the coins chosen. Observe students as they play.

FUNNY MONEY
(p. 52)

1. 8

2. 4; Q, D, N, P

3. 2; N, P

4. 6; 2 Q, D, 3 P

5. 8¢; N, 3 P

6. 65¢; 2 Q, D, N

7. Sample answer: $\frac{1}{4}$ of the coins could be worth 50¢ (as two quarters), while $\frac{1}{2}$ of the coins could be worth 13¢ (as 1 dime and 3 pennies); Kelly is right.

FRACTIONS AND AGES
(p. 53)

Answers will vary; check students' charts.

FRACTION ROLLERS, PART 2
(p. 54)

Greatest and least mixed numbers will vary depending on the numbers rolled.

NOTING FRACTIONS
(p. 56)

1. $\frac{1}{2} + \frac{1}{4} = \frac{3}{4}$

2. $\frac{1}{2} + \frac{1}{4} + \frac{2}{8} = 1$

3. $\frac{1}{4} + \frac{1}{8} + \frac{2}{16} = \frac{1}{2}$

4. $1 + 1 + \frac{1}{4} + \frac{1}{8} = 2\frac{3}{8}$

5. $\frac{1}{2} + \frac{1}{4} + \frac{1}{4} + \frac{1}{4} + \frac{1}{8} = 1\frac{3}{8}$

6. $1 + \frac{1}{16} + \frac{2}{16} = 1\frac{3}{16}$

7. $\frac{1}{2} + \frac{1}{8} + 1 + \frac{1}{4} + \frac{1}{8} = 2$

8. $\frac{2}{8} + \frac{2}{16} + \frac{1}{8} + \frac{2}{16} = \frac{5}{8}$

9. ♩ + ♩ + ♩ + ♪ = $\frac{7}{8}$

10. ♩ + ♩ + ♫ = 1

11. 𝅝 + 𝅝 + ♩ + ♫ = $2\frac{5}{8}$

12. ♩ + ♩ + 𝅝 + ♩ + ♩ = $2\frac{3}{4}$

13. 𝅝 + 𝅝 + ♩ + ♩ + ♪ + ♪ = $2\frac{11}{16}$

A HEAD FOR FRACTIONS
(p. 57)

Observe students as they play.

FRACTIONS IN ANCIENT EGYPT
(p. 59)

1. 🜔 2. ⊃ or ⊃ 3. 🜖

4. 👁 + 👁 5. 👁 + 👁

6. 🜖 + 👁

FLICKER FRACTION SUMS
(p. 61)

Answers will vary; observe students as they play.

FRACTION ADD-UP
(p. 64)

Answers will vary. Check students' work.

SUMS ON A ROLL
(p. 65)

Answers will vary; observe students as they play.

FRACTION PATH PUZZLES
(p. 67)

Path 1: +, −

Path 2: +, +, −

Path 3: +, −, +, −

FRACTION MAGIC FIGURES
(p. 69)

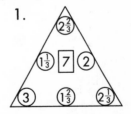

2.

$2\frac{2}{3}$	$\frac{1}{3}$	2
1	$1\frac{2}{3}$	$2\frac{1}{3}$
$1\frac{1}{3}$	3	$\frac{2}{3}$

3.

12	$4\frac{1}{2}$	6
$1\frac{1}{2}$	$7\frac{1}{2}$	$13\frac{1}{2}$
9	$10\frac{1}{2}$	3

4.

2	$\frac{1}{4}$	$\frac{3}{8}$	$1\frac{15}{24}$
$\frac{5}{8}$	$1\frac{3}{8}$	$1\frac{1}{4}$	1
$\frac{9}{8}$	$\frac{7}{8}$	$\frac{3}{4}$	$1\frac{1}{2}$
$\frac{1}{2}$	$1\frac{3}{4}$	$1\frac{7}{8}$	$\frac{1}{8}$

PARTS OF PARTS

(p. 71)

1. $\frac{1}{8}$
2. $\frac{1}{8}$
3. $\frac{3}{8}$
4. $\frac{1}{6}$
5. $\frac{1}{6}$
6. $\frac{2}{6}$ or $\frac{1}{3}$

ROLL, ROUND, AND RECORD

(p. 73)

Answers will vary; observe students as they play.

FRACTIONS AND CALORIES

(p. 75)

1. 1 hour
2. 15 minutes or $\frac{1}{4}$ hour
3. $\frac{2}{5}$ of an hour, or 24 minutes
4. She can bike or swim for about a half hour.
5. $\frac{350}{400}$, or $\frac{7}{8}$ of the calories
6. $1\frac{1}{2}$ cups of ice cream

FRACTION STORIES

(p. 76)

Answers will vary; check students' stories.

FRACTION SCAVENGER HUNT

(p. 78)

Solutions will vary; check students' scavenger hunt sheets.

ONLY ONE-THIRD AGREED THAT...

(p. 79)

Surveys and analyses will vary; check students' work.

FRACTIONS EVERY DAY

(p. 80)

Surveys and analyses will vary; check students' work.